PLACES

OF

INVENTiON

A Companion to the Exhibition at the
SMITHSONIAN'S NATIONAL MUSEUM OF AMERICAN HISTORY

EDITED BY
Arthur P. Molella and Anna Karvellas

WASHINGTON, D.C.
2015

Published by
SMITHSONIAN INSTITUTION SCHOLARLY PRESS
P.O. Box 37012, MRC 957
Washington, D.C. 20013-7012
www.scholarlypress.si.edu

Front cover images: (top) detail of first production planar IC, 1960, Fairchild Semiconductor. Courtesy of the Computer History Museum, Mountain View, California, Image #500004791; (middle) distillery of Corning and Company, Peoria, Illinois, about 1880–1919, Peoria Historical Society Collection, Bradley University Library; (bottom) detail of "Bustin' Loose" record by Chuck Brown and the Soul Searchers, 1978, with grease marks made by Grandmaster Flash, © 2014 Smithsonian Institution; photo by Richard Strauss. Courtesy of National Museum of American History.

Back cover images: (top) detail of Colt's east armory workers about 1900, PG 460, Colt's Patent Fire Arms Manufacturing Company collection, State Archives, Connecticut State Library; (middle) detail of researchers gathered on beanbags in the Computer Science Laboratory's Commons (circa 1980s), © parc, A Xerox Company; (bottom) detail of Earl Bakken at work in the Medtronic garage in Minneapolis about 1955, Courtesy of Medtronic, Inc.

Library of Congress Cataloging-in-Publication Data:

Places of Invention (2015 : Washington, D.C.)
 Places of invention / edited by Arthur P. Molella and Anna Karvellas.
  pages cm
 Accompanies the Places of Invention Exhibition at Smithsonian's National Museum of American History, 2015.
 Includes bibliographical references and index.
 ISBN 978-1-935623-68-7 (hardcover : alk. paper) — ISBN 978-1-935623-69-4 (ebook) 1. Inventions—United States—History—Exhibitions. 2. Industrial productivity centers—United States—Exhibitions. 3. Technology and state—United States—Exhibitions. 4. Industrial location—United States—Exhibitions. 5. Diffusion of innovations—Exhibitions. I. Molella, Arthur P., 1944– editor. II. Karvellas, Anna, editor. III. Title.
 T21.P57 2015
 609.73—dc23

        2014042368

Printed in the United States of America

⊗ The paper used in this publication meets the minimum requirements of the American National Standard for Permanence of Paper for Printed Library Materials Z39.48–1992.

CONTENTS

FOREWORD   v
Robert E. Simon Jr.

MUSEUM DIRECTOR'S NOTE   vii
John L. Gray

ACKNOWLEDGMENTS   viii

INTRODUCTION   1
Arthur P. Molella and Anna Karvellas

SILICON VALLEY (1970s–80s)   14
Eric S. Hintz | *Suburban Garage Hackers + Lab Researchers = Personal Computing*

BRONX (1970s)   42
Laurel Fritzsch | *Neighborhood Streets Create New Beats*

PLACES OF INVENTiON AFFILIATES PROJECT   66
Anna Karvellas | Dispatch from Greater Seattle | *The Rise of Seattle's Gaming Industry (1980s–Today)*

MEDICAL ALLEY (1950s)   86
Monica M. Smith | *Tight-Knit Community of Tinkerers Keeps Hearts Ticking*

HARTFORD (Late 1800s)   110
Eric S. Hintz | *Factory Town Puts the Pieces Together in Explosive New Ways*

PLACES OF INVENTiON AFFILIATES PROJECT   136
Anna Karvellas | Dispatch from Peoria | *Mass Production of Penicillin (1940s)*

HOLLYWOOD (1930s)   156
Joyce Bedi | *Follow the Rainbow to the Movies' Golden Age*

FORT COLLINS (2010s)   180
Joyce Bedi | *Campus and City Combine Their Energies for a Greener Planet*

PLACES OF INVENTiON AFFILIATES PROJECT   202
Anna Karvellas | Dispatch from Pittsburgh | *Jazz Innovation (1920s–70s)*

WHAT'S NEXT? MUSEUMS AND INNOVATION   222
Lorraine McConaghy

NOTES   231
BIBLIOGRAPHY   260
CONTRIBUTORS   291
INDEX   294

FOREWORD

This book and its companion exhibition explore the idea that most invention does not spring spontaneously from a single brain, but is fostered by the fertile soil of a particular place. Such "places of invention" can be identified by their geography, society, and creative culture. They make the most of natural and manmade resources and provide shared spaces for the exchange of ideas. The invention itself can be an object, such as a machine or device, or even a formula or a melody. It also can be a new or improved way of doing things, for instance, planning a new community.

My connection with the "new town" of Reston started in 1961 with the acquisition of 6,750 acres of land in Virginia located between Washington, D.C., and what is now Washington Dulles International Airport. I was involved in preparing its original plan and social programs and assembled a wonderful team of planners, architects, engineers, builders, realtors, educators, and sociologists to implement the program and bring it to market. At the time, we could not predict the rise of the Dulles Technology Corridor, but studies did forecast the substantial growth that would come to the Washington metropolitan region. We knew that Reston's proximity to Dulles Airport and the planned Dulles Access Road would provide key links between Reston, Washington, and the world that would heighten its appeal to prospective residents.

We wanted to do more for these residents than create another bedroom community. We wanted to create a place where people could live, work, and play—to create a walkable community with shared open spaces where people of all ages and backgrounds would gather. In 1965, the laudatory reception following the opening ceremonies, astonishingly international as well as national, indicated that the Reston team had, indeed, produced a new and improved way of building a community in the United States. In our plans were many of the concepts examined in this book; concepts which are still part of the ongoing discussion about the creation of innovation districts.

It has been said that Reston's influence on urban planning extends to many countries. As an indication of this, visitations from abroad started up almost immediately. In 1966, a Japanese crew arrived with a reporter, a translator, and a sound truck. Since then, a steady flow of visitors has continued. In the last five years signatures from fifteen different countries coming from five continents have appeared in the Reston Museum's guest book.

Of course, I am always happy to express the debt that Reston's original plan owes to my early influences:

- Separation of pedestrians from vehicles (Reston has nineteen underpasses and one overpass). The use of underpasses and overpasses can be traced to Leonardo da Vinci, who incorporated these elements into his designs for the suburbs of Milan, at the request of the Duke of Milan. I also was inspired by their use in Radburn, New Jersey, a planned community in which my father was involved.
- Lake Anne Plaza, the focal point for Reston's first arrivals, can be traced to plazas in most continents. I was particularly inspired by the plazas in Italian hill towns.
- The fifteen-story Heron House was inspired by the high-rise in Tapiola, Finland, outside of Helsinki. Heron House was built to create a "*here* here" in the Virginia boonies for Lake Anne Village Center. Without Heron House, Reston's debut might have suffered the same criticism that Gertrude Stein tagged to Oakland, California, complaining that there was no "*there* there."
- Sculpture for the delight of grown people and their offspring was a part of Reston's very beginnings. It was inspired by mature urban settings around the world.
- The idea for amenities such as walking-bike trails, a community center, ball fields, a swimming pool, and tennis courts came from Radburn. This concept was much ahead of its time.
- Lake Anne's fountain is a miniature of Lake Geneva's fountain.
- Atop Lake Anne Plaza's stores, townhouses had front yards with trees, shrubs, flowers, and grass. These townhouses were inspired by those in San Francisco's Embarcadero.

I have been prone to say that there is nothing new in Reston except, possibly, the collection of features that I had admired in my travels and personal experiences. Perhaps this collection, motivated by a desire to build COMMUNITY and to create another way of living, is itself an invention designed to foster the invention of others. Now more than 50 years old, Reston's story shares many characteristics with the places of invention explored in this book. One hopes that these stories inspire readers to make their own collections and think about ways to create places of invention where they live.

ROBERT E. SIMON JR.
Reston, Virginia

MUSEUM DIRECTOR'S NOTE

Innovation is at the heart of the indomitable American spirit. The National Museum of American History plays a critical role in inspiring the next generation of inventors, innovators, and entrepreneurs. We introduce our visitors to the stories that define America through hands-on exhibitions, programs, and performances. This museum tells America's whole story, in all its diversity, complexity, and possibility.

Places of Invention, in the Lemelson Hall of Invention and Innovation, is a signature exhibition that establishes the innovation theme on the first floor of the museum's west wing. Using six fascinating case studies and a network of online stories, *Places of Invention* examines the people, places, and circumstances that came together in the pursuit of something new. The first-floor exhibitions, learning places, and performance spaces that accompany *Places of Invention* offer visitors multiple ways to explore American innovation and the optimistic spirit, game-changing ideas, and willingness to fail and try again that are part of our collective story. These are stories about the role of invention and innovation in Americans' quest to move beyond the present and to shape the future.

The National Museum of American History is extremely proud to be the home of the Lemelson Center for the Study of Invention and Innovation, which has led the study of invention and innovation at the Smithsonian since 1995. Through its interdisciplinary and inspirational exhibitions and programs, publications, and partnerships—of which *Places of Invention* and this volume are a key part—the Lemelson Center continues to thoughtfully reveal the ingenuity, creativity, and inventiveness that are all hallmarks of the American character.

JOHN L. GRAY
Elizabeth MacMillan Director
National Museum of American History

ACKNOWLEDGMENTS

The authors thank the following individuals who provided advice and assistance with the *Places of Invention* book.

Silicon Valley: David Allison, Peggy Kidwell, Hal Wallace, and Nance Briscoe, National Museum of American History; Henry Lowood, Stanford University Library; Karen Kroslowitz and Sara Lott, Computer History Museum; Susan Kare and Heidi Nilsen, Susan Kare User Interface Graphics; Curt Carlson and Alice Resnick, Stanford Research Institute.

Bronx: Mark Katz, University of North Carolina; Christie Z-Pabon, Tools of War grassroots Hip Hop; Eric Jentsch and Stacey Kluck, National Museum of American History; Martha Diaz, NYU Hip-Hop Education Center; William E. Smith, Howard University; Carlos "Mare 139" Rodriguez; DXT.

Seattle: Tara McCauley, Leonard Garfield, Julia Swan, and Lorraine McConaghy, Museum of History & Industry; Helen Divjak; Sean Vesce, 20after1; Ed Fries; Megan Gaiser, Contagious Creativity; Richard Garfield; Jerry Holkins and Mike Krahulik, *Penny Arcade;* Andrew Perti, Seattle Interactive Media Museum; Kim Swift; Paul Thelen, Big Fish Games; Adam Wygle, Bootstrapper Studios.

Medical Alley: David Rhees, The Bakken Museum; Judy Chelnick, National Museum of American History; Lois Hendrickson and Dominique Tobbell, University of Minnesota; Manny Villafaña, Kips Bay Medical; Melissa Engelhart and Kristen Weaver, Medtronic, Inc.

Hartford: Roger White, Peter Liebhold, and David Miller, National Museum of American History; Dave Corrigan and Dean Nelson, Museum of Connecticut History; Christine Pittsley, Connecticut State Library; Rich Malley and Sierra Dixon, Connecticut Historical Society; Tom Condon, *Hartford Courant*; Bill Hosley, Terra Firma Northeast; Patti Philippon, Mark Twain House and Museum; Elizabeth Normen, *Connecticut Explored* magazine; Jackie Mandyck, iQuilt Partnership; Jelle Zeilinga de Boer, Wesleyan University.

Peoria: Diane Wendt, National Museum of American History; Kristan McKinsey, Ann Schmitt, Kate Neumiller Schureman, Renae Stenger, Cathie Neumiller, Jim Dwyer, and Zac Zetterberg,

Peoria Riverfront Museum; Mark Berhow, Nancy Davison, Pat Dowd, Kate O'Hara, and Stephen Peterson, U.S. Department of Agriculture's National Center for Agricultural Utilization Research; Chris Coulter and Robert Killion, Peoria Historical Society; Peter Johnsen; Liz Bloodworth.

Hollywood: Dwight Blocker Bowers, Ryan Lintelman, Franklin Robinson, Maria Elpert, Katherine Ott, and Shannon Perich, National Museum of American History; Jeffrey Flannery, Mike Mashon, Library of Congress; Bruce Torrence, HollywoodPhotographs.com; Nancy Kauffmann, James Layton, and Todd Gustavson, George Eastman House; Martin Hart, American WideScreen Museum; Robert Finley Doane and Deborah Douglas, MIT Museum; Julio Vera, Academy Museum of Motion Pictures; Mitchell Block, Direct Cinema Limited, Inc.; Thomas R. Sito, University of Southern California School of Cinematic Arts; Terri Garst, Los Angeles Public Library; Michael Hardy, Smithsonian Institution Libraries.

Fort Collins: Susan Evans and Hal Wallace, National Museum of American History; Bryan Willson, Morgan DeFoort, Justin Discoe, Wendy Hartzell, and Kate Heidkamp, CSU Engines and Energy Conversion Lab; Amy Prieto and Katie Hoffner, Prieto Battery; Judy Dorsey, Jessie Beyer, Kelly Flanigan, and Kirsten Savage, Brendle Group; Ed VanDyne, VanDyne SuperTurbo, Inc.; Sunil Cherian, Julie Zinn, and Nathan Howard, Spirae, Inc.; Kim Jordan, Jordanna Barrack, Andi Rose, and Bryan Simpson, New Belgium Brewing Co.; Tim Reeser and Bernita Kelley, CSU Ventures; Cheryl Donaldson, Annette Geiselman, and Lesley Drayton Struc, Fort Collins Museum of Discovery; Wade Troxell, Fort Collins City Council; Kortny Rolston, John Eisele, Mac McGoldrick, Joe A. Mendoza, and Joe Vasos, Colorado State University; Vicky Lopez-Terrill, Archives and Special Collections, CSU Libraries.

Pittsburgh: Ken Kimery and Wendy Shay, National Museum of American History; Samuel W. Black, Mariruth Leftwich, Kate Lukaszewicz, Alexis Smith Macklin, Brady Smith, Sandra Smith, and Matthew Strauss, Senator John Heinz History Center; Marty Ashby and Renée Govanucci, MCG Jazz; Dan Holland, Young Preservationists Association of Pittsburgh; Laurel Mitchell, Carnegie Museum of Art; Dr. Nelson Harrison, Pittsburgh Jazz Network.

Lemelson Center Team: Special thanks to Maggie Dennis and Amanda Murray; Jeff Brodie, Michelle DelCarlo, Tricia Edwards, Tanya Garner, Chris J. Gauthier, Claudine Klose, Steve Madewell, Alison Oswald, Tim Pula, William Reynolds, Matt Ringelstetter, Christopher White, Kate Wiley; and the Lemelson Center Advisory Committee.

POI **Exhibition Advisory Committee**: Leslie Berlin, Silicon Valley Archives, Stanford University; Paul Israel, Thomas A. Edison Papers, Rutgers University; Gretchen Jennings, *The Exhibitionist* Journal of the National Association for Museum Exhibition; John Kenny; Stuart W. Leslie, Johns Hopkins University; Peter Liebhold, National Museum of American History; Chip Lindsey, Science Works; Lorraine McConaghy, Museum of History & Industry; Hooley McLaughlin, Ontario Science Centre; Wendy Pollock; Kate Roberts, Minnesota Historical Society; Tom Simons, Joslyn Art Museum; Jim Spadaccini, Ideum; Heather Toomey Zimmerman, Pennsylvania State University.

Additional Advisors and Project Partners: Frances Dispenzirie, Jaclyn Nash, Rosemary Phillips, Richard Strauss, Hugh Talman, Kari Fantasia, Mike Johnson, and Maggie Webster, National Museum of American History; Jennifer Brundage, Elizabeth Bugbee, Harold Closter, Alma Douglas, Aaron Glavas, Laura Hansen, and Caroline Mah, Smithsonian Affiliations; Alphonse DeSena, National Science Foundation; Maria Scaler, Smithsonian Office of Sponsored Projects; Martha Davidson; Carol F. Inman; Roto Group LLC; Randi Korn & Associates, Inc.; all Phase One and Phase Two *Places of Invention* Affiliate museums and their community partners; Ginger Strader, Smithsonian Institution Scholarly Press; and Piper Wallis.

Exhibition Funders: National Science Foundation; Intel Corporation; Lemelson Foundation; and Charles Stark Draper Laboratory.

INTRODUCTION

Arthur P. Molella and Anna Karvellas

Does place really matter for invention? Predictions that cyberspace will supplant physical communities and that spatial proximity will become irrelevant have been around for a while. If you live in Kansas City but have a meeting in New York City, no worries—just stay at home and fire up Skype. If these predictions are true, however, we have to ask why leading computer and Internet companies like Google and Apple not only invest in corporate campuses but also insist on clustering them in Silicon Valley. Those giants of the digital realm are revealing something important about the modern world of innovation: Yes, place still matters. Expectations run exceptionally high for technology regions—networks, clusters, hot spots, or high-tech corridors—to perform as forces for national, technological, and economic regeneration. A constant dilemma today—in the United States and abroad—is how to create and sustain innovative environments, whether on the personal scale of individual labs or at the regional level.[1]

With all the contemporary buzz around innovation, the ideas behind *Places of Invention* and its accompanying exhibition at the Smithsonian's National Museum of American History could not be more timely. Building on twenty years of research by the Lemelson Center, this book and its companion exhibition ask, What is it about some places that sparks invention and innovation? Is it simply being at the right place at the right time, or is it more than that? How does *place*—whether physical, social, or cultural—support, constrain, and shape innovation? Why does invention flourish in one spot but struggle in another, even very similar location? In short, Why there? Why then?

A CASE STUDY APPROACH

To address these questions, the *Places of Invention* book and exhibition are framed around six historical case studies that take readers and museum visitors on a journey through time and place to discover the stories of people who lived, worked, played, collaborated, adapted, took risks, solved problems, and sometimes failed—all in the pursuit of creating something new. These examples were informed by the Lemelson Center's research and the richness of the museum's archival and artifact collections and were selected to represent an intriguing array of people, places, time periods, and technologies. Although the case studies are based in the United States, we believe that they have much in common with international developments.

- Silicon Valley, California, 1970s–80s: Collaboration was essential to the technological advances behind the birth of the personal computer and Silicon Valley's enduring fame.
- Bronx, New York, 1970s: Early hip-hop pioneers and their technical innovations were characterized by adaptability and flexibility.
- Medical Alley, Minnesota, 1950s: Risk taking and collaboration among heart surgeons and engineers helped the Twin Cities earn their reputation as a major center for medical innovations.
- Hartford, Connecticut, late 1800s: Networks and knowledge sharing led to improvements in precision manufacturing, from guns to sewing machines to bicycles.
- Hollywood, California, 1930s: Initiative and creativity drove Hollywood in its golden age with inventions in color technologies for motion pictures.
- Fort Collins, Colorado, 2010s: Critical thinking and problem-solving skills are being applied to sustainable clean energy innovations.

Our case studies do not follow any special overall sequence, either chronologically or in terms of subject area. In both the book and the exhibition, we wanted to encourage visitors to make comparisons between inventive places, without forcing upon them a narrative of chronological or technological progress, which cannot be justified historically. The *Places of Invention* exhibition is designed to encourage these kinds of connections. The six case studies are arranged around a central intersection called "the Hub." Low walls and shared sight lines allow visitors to follow their interests and create their own paths through the stories. Modern exhibition practice encourages such opportunities for visitor choice. Each setting is a unit unto itself, but visitors will also recognize discernible patterns *across* our case studies in time and lines of technological development. That said, we did arrange for a couple of juxtapositions along the circle of places. We thought putting Silicon Valley next to the Bronx, both set in the 1970s, was thought provoking and raised interesting questions about the conditions for inventive activity. We placed Fort Collins, our only contemporary case study, near the exit in order to have visitors leave with a look to the future (Figure 1).

We believe that history offers valuable lessons for understanding how new cultures of innovation develop, from the personal level to the local, regional, or national scale and beyond. Although

FIGURE 1

Places of Invention exhibition floor plan schematic by Roto Group LLC. © Smithsonian Institution; courtesy of the Lemelson Center for the Study of Invention and Innovation.

they may feel very recent to us, modern places of special scientific, technological, and cultural ferment have many historical precedents. As historian of technology Jennifer Light has observed, "Scholars have identified changes in the inventive process from the late 19th through the 20th centuries, suggesting the value of taking a historical approach to assess even contemporary innovation practices."[2] Our case studies, then, help us understand a complex historical, yet still evolving, phenomenon.

Some of the most significant investigations of innovative places involve not only start-up developments but also how to sustain an innovative culture after initial success.[3] We have observed that creative places go through life cycles and manifest peaks and valleys of productivity. We know much more today about sustaining creative cultures, and the motivations for doing so are stronger than ever before. Looking forward, we expect and firmly hope that inventive places will become the norm rather than the exception, with the prospect of long and healthy life spans.

Innovation is bubbling up all around us, and we could have easily found six other equally suitable cases. Rather than a scientific sample, our selections were shaped by the needs and approach of the exhibition: We sought diversity of people and communities; variety in time and space; familiar, accessible inventions that are available in our museum collections; and, last but not least, good stories. To increase the range and scope of these stories, we also conceived of an interactive map at the center of the exhibition that would allow us to present crowdsourced visitor stories and video "case studies" curated with Smithsonian Affiliate museums.[4]

EXTENDING SCHOLARSHIP

The literature about invention, innovation, and place is extensive and growing. Innovative people, labs, companies, and regions have attracted the attention of a multidisciplinary array of scholars, including historians, geographers, economists, urbanists, sociologists, and business experts. For example, business historian Margaret Graham studied the climate for research and development within such corporations as Corning and Alcoa, and David Hounshell and John K. Smith documented the rise of DuPont's iconic research and development lab. Journalist Jon Gertner has studied the innovative culture of Bell Labs.[5] Federal funding, especially from the Department of Defense, has been crucial to the development of high-tech regions, as Christophe Lécuyer has shown for Silicon Valley. In the world of government-funded labs, Lillian Hoddeson, Adrienne Kolb, and Catherine Westfall have documented the Fermi National Accelerator Laboratory (Fermilab) and the practice of what they call "megascience."[6] Historian Stuart W. Leslie's interests range from laboratory architecture to the rise and fall of high-tech regions.[7] Regionalism as a corporate strategy is the theme of several studies by Harvard University's Michael E. Porter, who finds that "paradoxically, the enduring competitive advantages in a global economy lie increasingly in local things."[8]

Maryann Feldman has explored the geography of innovation, showing how local industrial clusters enable the free flow of knowledge, skills, people, and, not least, money between institutions—be they universities, governments, or corporations.[9] In these regional nodes, informal communication and social gatherings among individual actors and institutions are crucial. In a colorful 1983 article in *Esquire* about Intel's Robert Noyce, writer Tom Wolfe describes meetings among engineers, start-up founders, and venture capitalists at Silicon Valley watering holes like the Wagon Wheel and Chez Yvonne.[10] In addition to Silicon Valley, *Places of Invention* documents equivalent gathering spots where creative people and resources came together, from Hollywood's Brown Derby and Fort Collins's New Belgium Brewery to city parks in the Bronx.[11]

Highly relevant to our themes are the seminal studies of political scientist and information expert AnnaLee Saxenian, who focuses on the interactions of firms within regions. In *Regional Advantage* and other books, Saxenian explores the cultural dynamics of industrial adoption in high-tech regional networks. She points out that although physical proximity is critical, just being neighbors is not enough: "Firms are embedded in a social and institutional setting that shapes, and is shaped by, their strategies and structures."[12] Economist Ann Markusen, who had previously explored the role of the military and defense corporations in the rise of high-tech regions, has now turned her attention to culture and "creative placemaking," focusing on the intersection between artistic culture and urban/regional development. Making the biggest splash in this area is urban studies expert Richard Florida, who links the new innovation economy to the rise of a "creative class," made up of people who create for a living—artists, scientists, inventors, engineers, architects, and entertainers, among others.[13] Collectively, these scholars, working across a range of disciplines and time periods, have established the importance of understanding the role of geography and local culture in fostering innovative places.

Many of the scholars cited above have shared their expertise with the Lemelson Center over the past two decades. Since its founding in 1995, the center has fostered the study and exploration of the role of place in invention and innovation, beginning with our inaugural New Perspectives on Invention and Innovation symposium on "The Inventor and the Innovative Society."[14] Speakers, including AnnaLee Saxenian and Stuart W. Leslie, examined the ways in which three innovative societies—Renaissance Italy, metropolitan New York in the late nineteenth century, and California's Silicon Valley in the latter twentieth century—nurtured and sustained the inventive impulse of such luminaries as Leonardo da Vinci, Thomas Edison, and Frederick Terman.

As the center evolved, we continued to research and document historical and contemporary inventors and innovators, study their invention processes, and examine the relevance of place and culture in their lives and careers. We visited inventors' workspaces; acquired sketches, models, and records; conducted oral histories; and documented environments using photography and video.

Our burgeoning research effort led to a conference on "Cultures of Innovation" at the National Museum of American History in 2005.[15] Participants addressed and debated global, cross-cultural, and interdisciplinary perspectives about the relationships among individuals, social structures, cultural practices, and technical, legal, and governmental infrastructures that support (or even discourage) invention and innovation.

These early investigations gave us a broad sense of the temporal and spatial dimensions of innovative places—a phenomenon deeply rooted in time and truly global in extent and impact. To enhance our understanding of the relationship between physical spaces and creativity, we convened an interdisciplinary group of scholars and practitioners in 2007 for the Lemelson Institute on Places of Invention. The Institute's findings offered insights into the qualities of physical space that are conducive to innovation, the ways that creative people shape the spaces in which they work, and the common creative features of inventive spaces and places—from the garages and basements of independent inventors to academic and government laboratories to cities, regions, and even cyberspace. These findings strongly inform this book and the exhibition:

- Places of invention that "work" share some common features, including flexibility, understated leadership, good communication, and a balance between individual and collaborative work.
- Communities, whether large or small, play an important role in shaping places of invention. Even the quintessential "lone inventor" is part of one or more groups and communities. Conversely, most creative groups have a leader, that charismatic person around whom teams form.
- Inventors and the many communities of which they are a part are affected by their social and intellectual networks, by changing forms of communication, and by the patent system. However, trying to create a new community of invention by replicating a successful model seldom succeeds.
- Creative spaces and places, from laboratories and institutions to cities and regions, go through life cycles of varying productivity.[16]

The Lemelson Institute findings also shaped our next study on the connections between place and invention. In 2008, we held a workshop on "Places of Invention: An Exploration of Best Practices for Documenting the Labs, Workshops, and Creative Spaces of America's Inventors."[17] To launch the discussions, the participants toured two spaces of invention in metropolitan Washington, D.C.: the home basement workshop and laboratory of Chuck Popenoe, inventor of the SmartBolt and founder of Stress Indicators, Inc., and the Maryland Technology Development Center, an incubator specializing in biotechnology startups.[18] These visits laid the foundation for conversations about the roles played by place, tools and materials, personal identity, professional networks, and community in the inventive process.

This new interpretation of innovative places emerged alongside 21st Century Skills, a complementary educational reform movement launched by the Partnership for 21st Century Skills.[19] Incorporating such cognitive and social skills as collaboration, adaptability, risk-taking, communication, creativity, and problem solving, the 21st Century Skills movement resonates with the findings of the Lemelson Institute and provides useful ways to see commonalities across *Places of Invention* case studies. Throughout the chapters in this book we examine these important inventive skills and behaviors and how they were manifested in each place. These same behaviors are also highlighted in "Skill Spots" within each case study in the exhibition.

We began to test ideas for the exhibition during 2009 in a small showcase exhibit and in our New Perspectives symposium on "Hot Spots of Invention: People, Places, and Spaces." This gathering of historians, practitioners, and a broad range of audiences explored what would become some of the *Places of Invention* exhibition case studies—Silicon Valley, California; Medical Alley, Minnesota; Hollywood, California; and Fort Collins, Colorado—as well as other stories that illustrated the powerful historical interplay of people, places, resources, and ideas in shaping inventors' work.

In 2010, our *Places of Invention* exhibition project received a generous grant from the National Science Foundation, allowing us to move ahead with content development, evaluation strategies, and design for an interactive and highly engaging exhibition for the public. We knew that *Places of Invention* was not standard exhibition fare, even for science and technology museums. Indeed, our initial evaluations revealed that the public could not always connect our ideas about place to their own lives and personal inventiveness. To many, inventors were a mysterious "other," and we realized we had to adjust our approach. We learned that if we referred not just to inventors but to inventive or creative people, we seemed to hit a resonant note with our audiences. Also, we confirmed the importance of defining key terms. We focused our definition of *place* at the city and regional level. *Spaces* within inventive places include labs, workshops, kitchens, garages, and so on. *Places of Invention* embraces both *invention* and *innovation*, the distinctions between which are widely debated. In fact, the general public often conflates the two terms. A standard definition is that invention involves novelty, the creation of a new thing or process based on new ideas; innovation involves the introduction of an invention in some form or fashion into society and general use.[20]

ECOSYSTEMS OF INVENTION

Places of invention are more than statistical data, spatial processes, and maps. Ultimately, they come down to individual passions and sparks of genius. As we have seen, a large body of literature on innovation at the company, regional, and national levels exists. Less well understood, however, are the physical and social spaces of individual inventors. Although artists' studios and scientists' labs have received serious academic study, the relationship between inventors with their work spaces has received far less attention.[21] This lack of attention may be because inventors' work environments, especially those of independent inventors, are hard to pigeonhole: they can be not only institutional labs but also dens, kitchens, basements, garages, and even barns.[22] Also needing closer study are the relations between inventors and their local networks from the standpoint of the inventor, that is, from the bottom up, rather than from the top down. Connecting personal narratives with institutional and local histories, *Places of Invention* aims to close the gap between the terrain of individual inventors, their personal and professional networks, and the geography of the technology region.

"Innovative activity tends to cluster spatially in what is known as the 'Silicon Valley' phenomenon," writes Feldman,[23] and today's civic, government, and business leaders frequently ask, "How can we replicate Silicon Valley? Is there a recipe for its success?"[24] It turns out that the answers to these questions are anything but straightforward. Spin-offs and replicated regions have rarely been successful (Leslie has documented some conspicuous failures).[25] Given the more fluid model of innovation today, where so much depends on serendipity and the mysterious chemistry of human interaction and place, it is far more difficult—probably impossible—to recreate Silicon Valley than to explain it in hindsight or even to sustain it. But it is the ultimate mystery of human creativity that makes the places of invention story so compelling. For example, even as Silicon Valley was acquiring its famous nickname in the early 1970s, a distant location on the East Coast of the United States witnessed its own eruption of innovation, but starting from a drastically different situation.

The Bronx, New York, was as economically destitute as Silicon Valley was rich. However, by the late 1970s, it had emerged as one of the nation's most creative hot spots. It was the birthplace of hip-hop, a revolutionary amalgam of technology, music, and culture that by the late 1980s became a staple of MTV and was on its way to becoming a billion-dollar global industry. In most ways, Silicon Valley and the Bronx could not have been more different, and yet as inventive places they had many features in common with each other, as well as with other places of invention documented in this volume. A successful place of invention is inherently a fertile blend of place, people, and invention. Using these broad crosscutting elements as a rough organizing device, *Places of Invention* teases out common themes from the six featured case studies. A few examples will suggest some of them.

Beginning with individual inventors, listen to the Bronx's DJ Grandmaster Flash, one of hip-hop's technological virtuosos:

> I was a scientist looking for something. Going inside hair dryers, and going inside washing machines and stereos and radios, whatever you plugged into the wall. . . . Although there was crazy violent things happening around me on Fox Street, I was in my own world, in my own room.[26]

In the quest for something new and different, inventors typically scavenge for tools and parts in their local environment. In creating the customized turntables and enormous amplifiers that made them local celebrities, aspiring DJs used whatever they could find at hand in their homes or in the streets. These found objects were the inventor's equivalent of the artist's palette. Out of the shards of everyday life, including parts from abandoned cars and electricity siphoned from lampposts, a new, pieced-together technology gave rise to a radically new musical form, itself defined by shards of sound gleaned and reassembled from multiple sources, musical and technological—hip-hop's trademark sampling, mixing, and scratching techniques. This inventing style, a kind of genius-level tinkering, fits the classic definition of bricolage: "construction or creation from a diverse range of materials or sources . . . often in found objects."[27] It also happens to be a style associated with musical modernism from Igor Stravinsky to Charlie Parker.[28] Inventing is often compared, and rightly so, to jazz improvisation.

The Bronx DJs and other hip-hop pioneers were prime examples of austerity serving as a spur to innovation and creative problem solving. In stark contrast, Steve Jobs came of age as an inventor in affluent, technology-rich northern California. However, Jobs and co-inventor Steve Wozniak were doing essentially the same thing as DJs in the Bronx, scrounging bits and pieces of electronics wherever they could find them—even "liberating" spare parts from their workplaces at Hewlett-Packard and Atari—and then reassembling them into the new conception of a personalized computer that they believed would upend technology and society. Similarly, Medtronic founder Earl Bakken improvised pieces of electronics into the first wearable, external cardiac pacemaker. It started as a patched-together device, or kludge, but evolved into a sleek miniaturized implantable device that looks like it was always meant to be. In short, the whole had become more than the sum of the parts, the shape of something completely new. Such transformations can border on the magical: Silicon Valley firms turn sand into semiconductors; Hollywood innovators convert the silver screen into the rainbow. Although we hear today about the recent rise of the creative class, it clearly made itself felt earlier in the century and not only in 1930s Hollywood.

These visionaries exemplified divergent thinking, a creative thought process based on "flow" and spontaneity—"Think different," in the words of the Apple company slogan.[29] Again, place entered heavily into the equation. "I was in my own world, in my own room," Grandmaster Flash tells us. For most inventors, moments of solitude, in some safe personal space, can be crucial to creative breakthroughs. David Kelley, inventor and founder of the IDEO design firm, put up a whiteboard in his shower so he could jot down ideas immediately. The garages of inventors like Bakken in Minneapolis or Jobs and Wozniak in Los Altos were a haven not only for tinkering but also for insight and reflection. However, none of them worked in a solitary world. They and many other well-known inventors, going back to Alexander Graham Bell and Thomas Edison, relied on teamwork, collaboration, and competition to spur their inventiveness.

Sociologists find that charismatic leaders such as Edison and Jobs play a crucial role in developing creative spaces, attracting inventive people, and getting funding. More important, they connect individual practitioners with the larger scene of social and collaborative networks that define technology regions. Such a shaping role was played by Samuel Colt for Hartford, Bryan Willson for Fort Collins, and Dr. C. Walton Lillehei for Medical Alley. A symbiosis develops between the individual and the region: Community brings a wealth of resources to the innovator, whether independent or institutionally based. Individual innovators in turn contribute their skills, knowledge, and creativity to the larger ecosystem of invention. If a tipping point is reached, a node can blossom into a full-blown technology region.[30]

This exchange and sense of community (although not always comity because disagreement often eventuates into profound change) are necessary ingredients in a rich culture of innovation, as appreciated by scholars like Markusen, Saxenian, and Florida. Such a culture potentially produces not only breakthroughs in technology but also new forms of art, architecture, music, or film that often shake up the status quo. In his foreword to this volume, Robert E. Simon Jr., the eponymous founder of Reston, Virginia, makes the point about the importance of community culture from a position of unequaled experience and success: "Community, to the extent that a developer can engender it, that is the most important thing you can do," he said in a recent interview with the Lemelson Center.[31]

THE *PLACES OF INVENTiON* AFFILIATES PROJECT AND INTERACTIVE MAP: TAKING IT TO THE NATION AND THE WORLD

- What if we created *Places of Invention* "learning labs" that would take these ideas further—out into the nation where local museums, community partners, and the general public could document their places of invention?
- What methods and platforms could we utilize to show how invention can be used as a transformative lens for understanding local history, cultivating creativity, and engaging communities?
- Could our project inspire a dialogue about innovation—not a one-way "official" narrative—that would allow the Smithsonian and its partners to learn from each other and the world at large?

The Lemelson Center was eager to explore these questions, as was the National Science Foundation, whose grant also supported the launch of the *Places of Invention* Affiliates Project. At its heart, this project was designed to present a new model for the collaborative creation of exhibition content and to bring Smithsonian resources to those who might never visit one of our institutions.

Working with our colleagues at the Smithsonian Affiliations office, we initially selected six regionally diverse Affiliate institutions—out of a network of nearly 200—to research and document local invention and innovation. In a second round, we recruited an additional six Affiliates for a total of twelve teams. Affiliates received funding to work with community partners to develop and present content related to *Places of Invention*. With Smithsonian training and guidance, teams produced deliverables ranging from public programs and oral histories to short videos for the centerpiece of the *Places of Invention* exhibition: a large interactive map encouraging visitors to explore, discuss, and contribute stories about innovative communities.

For the exhibition opening, the map was populated with Affiliate videos showcasing how people, resources, and geography had combined to make each community a place of invention. Videos focused on historic and/or contemporary invention and examined cycles of innovation driven by natural resources, evolving infrastructure, historic events, and the people who came together in a particular time and place to do something new. They follow geographic and geologic advantage; great American rivers such as the Illinois, Ohio, Monongahela, Allegheny, and Merrimac; westward expansion and a growing network of railroads, canals, and highways; industrialization; Prohibition; and World War II. They chronicle skill sets and communities: the gaming expertise that emerged from Microsoft's incubation of Greater Seattle technology talent; the computer-aided design industry that sprang from IBM's work on NASA's Apollo program in Huntsville, Alabama; and anticounterfeit innovations developed over two centuries of association between the U.S. Treasury and Crane & Company in Massachusetts' Berkshires.

The *Places of Invention* map will grow exponentially over time with in-gallery and online contributions crowdsourced from anywhere in the world, including but not limited to Smithsonian Affiliate museum galleries. This last factor is especially important: Using the interactive map as a local tool, Affiliates can create ever-increasing opportunities for discussion, debate, and content development. Have a story related to the video? Share it. Disagree with our take on the subject? Tell us why. These opportunities for involvement and contribution are especially welcome at museums trying to engage groups who have historically been marginalized from local narratives. Want to explore other places of invention? Search the stories for places with similar tags—you might be surprised how seemingly similar contexts can spark completely different kinds of innovation.

In developing local project "prototypes," Affiliates were encouraged to use the map, public programs, and other project tools to set up opportunities to explore invention and innovation for years to come. Some created exhibitions around the interactive map or used it to complement already existing museum work. Some launched *Places of Invention*–themed lecture series and family days or added *Places of Invention*–themed activities to preexisting programs.

We worked closely with public historian and *Places of Invention* project consultant Lorraine McConaghy, whose *Nearby History* program at Seattle's Museum of History & Industry directly inspired the Affiliates Project. For more than a dozen years, McConaghy gave workshops that provided members of the Greater Seattle community with the skills and resources needed to conduct historical research about their homes, neighborhoods, and families. McConaghy's public history work continues to push the envelope and inspire, most recently through projects with the

Museum of History & Industry's Bezos Center for Innovation, the Washington State Historical Society, and Ford's Theatre in Washington, D.C.

To kick off both phases of the *Places of Invention* Affiliates Project, the Lemelson Center brought Affiliates and their community partners to the National Museum of American History for intensive training workshops. Joining them at each event were other members of the *Places of Invention* collaborative community, including Smithsonian Affiliations staff and National Science Foundation program officer Alphonse DeSena. The heart of the training was a series of presentations, training activities, and brainstorming sessions led by McConaghy. Collectively called "A Toolkit to Document Your Place of Invention," these resources were designed to give participants different strategies for finding an interpretive pathway to their *Places of Invention* topic. "Let's be reflective," McConaghy advised. "Think about your topic as an opportunity to engage your community in invention and innovation, and inspire people to think of their own place as a place of invention."[32]

The twelve groups participating in the *Places of Invention* Affiliates Project include the following Smithsonian Affiliates and their community partners.

Phase One teams include:

- American Textile History Museum, with Lowell Telecommunications Corporation
  Lowell, Massachusetts
- Museum of History & Industry, with 20after1
  Greater Seattle, Washington
- Peoria Riverfront Museum, with the U.S. Department of Agriculture's National Center for Agricultural Utilization Research and the Peoria Historical Society
  Peoria, Illinois
- Senator John Heinz History Center, with MCG Jazz and the Young Preservationists Association of Pittsburgh
  Pittsburgh, Pennsylvania
- The Works: Ohio Center for History, Art, and Technology, with Owens Corning
  Newark, Ohio
- U.S. Space and Rocket Center, with the University of Alabama at Huntsville
  Huntsville, Alabama

Phase Two teams include:

- Berkshire Museum, with the Berkshire Historical Society
  Pittsfield, Massachusetts

- Conner Prairie Interactive History Park, with Howard County Historical Society and the Indiana Gas Boom Heritage Area
  Kokomo, Indiana
- The National World War II Museum, with the University of New Orleans
  New Orleans, Louisiana
- ScienceWorks Hands-on Museum, with the Southern Oregon Historical Society
  Ashland, Oregon
- Telluride Historical Museum, with the Pinhead Institute
  Telluride, Colorado
- Western Reserve Historical Society, with WVIZ/PBS & WCPN ideastream
  Cleveland, Ohio

The three "Affiliate Project Dispatches" in this book showcase *Places of Invention* work in Seattle, Washington; Peoria, Illinois; and Pittsburgh, Pennsylvania, to illustrate varying approaches to this project. Each dispatch provides a brief overview of the museum and its community, their reasons for selecting a particular place-based invention topic, what they learned by using *Places of Invention* as a lens for exploration, and what the outcomes have been to date. These pieces have a different tone and look than the rest of the book to convey a sense of place, to reflect the voices of the Affiliate teams and their communities, and to acknowledge that these are active explorations meant to live on in different iterations for years to come.

Some of the great privileges of this project have been to visit these and other places of invention; to tour museums and factories and labs and coffee shops and canals and salmon ladders and giant earthworks; to talk to people on the project and in the community about their role in past and present innovation; and to share in the surprise and enjoyment that come from making new discoveries and unlikely associations. We have asked what people think others should understand about their place of invention and why they think it is conducive to creativity and innovation. *Places of Invention* teaches that invention and innovation are about you and me and where we live and what we do to create any new or improved way of doing things. They are about the communities that cultivate this kind of experimentation and what and how and where we learn from each other.

As McConaghy writes in this volume's final essay, museums have the potential to initiate regenerative, optimistic cultures of innovation in their communities. In this context, they are more than relevant—they are *essential*. It is our hope that museums, historical societies, schools, business districts, and other state and local institutions can use *Places of Invention* as a model and embrace the idea that a better understanding of invention history—its cycles, hot spots, range of technologies, failures, and successes—can help inspire new cultures of innovation. We hope, too, that this book and exhibition will inspire individuals—regardless of age or background—to see themselves as inventive and active contributors to *their* places of invention.

SILICON VALLEY

Suburban Garage Hackers + Lab Researchers = Personal Computing

CALIFORNIA (1970s–80s)

Eric S. Hintz

Silicon Valley is perhaps today's most recognizable place of invention. This cluster of towns south of San Francisco, including Palo Alto, Menlo Park, Cupertino, Sunnyvale, Mountain View, Santa Clara, and San Jose, has been the epicenter for innovations in solid-state electronics, personal computing, networking, software, social media, and the venture capital sector that funds them. But California's Santa Clara Valley was not always known for high tech. This region, known earlier as the Valley of Heart's Delight, had once been an agricultural paradise, teeming with fruit orchards and canneries. Over time, its sunny weather, attractive suburbs, proximity to Stanford University, and casual but fiercely entrepreneurial business culture attracted talented people and new businesses to the region. A booming electronics industry emerged in the 1960s and inspired the new nickname, Silicon Valley, after the main element in integrated circuits. Then, in the 1970s and 1980s, the region nurtured the invention of the personal computer.

PLACE
From the Valley of Heart's Delight to Silicon Valley

The Santa Clara Valley owes much of its success to its geological and natural endowments. To the north, San Francisco Bay is one of the world's largest natural harbors and became the focal point of California's settlement in the eighteenth and nineteenth centuries. The valley itself sits at the southern end of San Francisco Bay and is ringed by the Santa Cruz Mountains to the west and the Diablo Range to the south and east (Figure 1). Over the millennia, erosion and runoff from these surrounding mountains endowed the valley with fertile soil. These mountains, especially the Santa Cruz range to the west, often block the Pacific coastal fog that is synonymous with San Francisco to the north, which means that the Santa Clara Valley is typically sunny with a relatively narrow

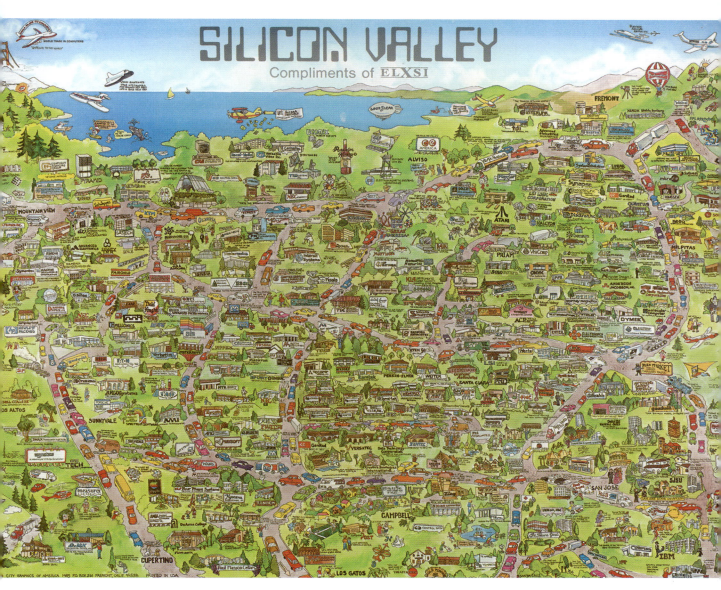

FIGURE 1
This poster map of Silicon Valley from 1983 shows the location of the region's high-tech firms and other area landmarks. *Silicon Valley Map* by City Graphics of America, courtesy of the Computer History Museum, Mountain View, California.

temperature range; for example, only 40°F separate San Jose's average low of 42°F in January and average high of 82°F in August. This absence of extremes, with little frost in the winter and few heat waves in the summer, prevented disastrous crops losses and ensured a steady and long growing season that would eventually make the valley one of California's most productive agricultural regions.[1]

The native Ohlone tribes were the valley's first inhabitants, and they lived undisturbed for thousands of years. Beginning in 1769, Spanish conquistadors, led by military governor Don

Gaspar de Portolá and Franciscan friar Junípero Serra, traveled north from Mexico along the so-called El Camino Real
(royal road) and established a series of 21 Catholic mission
churches along the California coast, including Mission
Santa Clara (1777) and Mission San José (1797). Vulnerabil-
ity to European diseases (e.g., the measles epidemic of 1808)
decimated the Ohlone and cleared the way for further settlement, which quickened considerably in the 1840s. During the
Mexican-American War, Commodore John D. Sloat claimed California for the United States on 7 July 1846. Two years later, in 1848, gold
was discovered in the Sierra Nevada foothills east of the Bay Area, changing California forever.
Thousands of forty-niners arrived by ship and wagon train during the next year, and San Francisco became an overnight boom town as the busiest port and largest commercial center west
of the Mississippi. Annexation and California statehood followed quickly in 1850, with San Jose
initially serving as the first state capital.[2]

Meanwhile, the Santa Clara Valley became an agricultural breadbasket that fed the onrush of
California settlers. Wheat production peaked in the 1880s and eventually gave way to fruits, nuts,
vegetables, and other types of produce. Accordingly, an 1896 description in *Santa Clara County
and Its Resources* described the region's "Fruitful Orchards," "Great Drying Establishments," and
"pioneer packers" and canners, which grew and processed the "Profitable Prune," the "Golden
Apricot," and the "Kingly Orange." Agriculture would dominate the region until after World War
II, when farmers increasingly sold their orchards to make way for new electronics firms and tract
homes for returning GIs (Figure 2).[3]

A crucial milestone in the valley's development was the founding of Stanford University. New
York native Leland Stanford (1824–1893) followed the Gold Rush west in the 1850s and made his
fortune, first as a merchant and supplier of mining equipment, then as one of the financiers of
the Central Pacific branch of the transcontinental railroad. During a family trip to Italy in 1884,
fifteen-year-old Leland Stanford Jr. contracted typhoid fever and died; the elder Stanford and his
wife, Jane Lathrop Stanford, built the university as a memorial to their late son. The couple do-
nated their 8,180 acre Palo Alto horse farm for the campus and established an unprecedented $20
million endowment; in 1891, the Leland Stanford Jr. University welcomed its first students. Over
time, it grew into a world-class institution, with a strong focus on practical training in science and
engineering.[4]

Stanford University was instrumental in cultivating the Santa Clara Valley's early electron-
ics industry. For example, in 1909 Stanford graduate Cyril Elwell convinced university president
David Starr Jordan and engineering professor C. D. Marx to finance his new venture, the Federal
Telegraph Company, to provide wireless telephone and telegraph services on the West Coast

> Silicon Valley is the only
> place on earth not trying
> to figure out how to
> become Silicon Valley.
>
> ETHERNET INVENTOR
> AND 3COM FOUNDER
> ROBERT METCALFE

Orchard and foothills, Santa Clara Valley, Cal.

FIGURE 2
A postcard highlighting the orchards and foothills of the Santa Clara Valley, nicknamed the Valley of Heart's Delight, circa 1900–1910. Courtesy History San José.

and later for the U.S. Navy. In 1932, Stanford graduate and Federal Telegraph Company veteran Charles Litton founded Litton Engineering to build glass lathes and vacuum tube–making machinery; later in 1945–46, he spun off Litton Industries, which built magnetron microwave emitters used in military radar systems. Litton was introduced to the microwave business after he supplied tube-making equipment to fellow Stanford graduate Russell Varian (and his brother Sigurd), who borrowed laboratory space from the Stanford Physics Department in 1937–38 to develop their klystron high-frequency microwave transceiver. Most famously, Stanford electrical engineering professor Frederick Terman encouraged former students William Hewlett and David Packard by helping them find some klystron work with the Varian brothers, part of which they conducted in Litton's Redwood City laboratory. Hewlett and Packard subsequently formed a partnership and built a new type of audio oscillator, working out of Litton's lab and a rented Palo Alto garage. They sold the oscillator to Walt Disney Studios to use for the movie *Fantasia* and founded the Hewlett-Packard Company in 1939. As these Stanford-trained inventors built their firms, they developed skills in design, product engineering, manufacturing, sales, and management. Thus, long before the valley's first semiconductor firms emerged in the 1950s and 1960s, the region's pioneering electronics firms had already developed the technical and managerial skills necessary to introduce and commercialize new innovations.[5]

> This is one of the most fertile of the many small valleys of the coast; its rich bottoms are filled with wheat-fields, and orchards, and vineyards, and alfalfa meadows.
>
> NATURALIST AND SIERRA CLUB FOUNDER JOHN MUIR DESCRIBING THE SANTA CLARA VALLEY

FIGURE 3
The Navy's USS *Macon* dirigible, moored at South Circle near the massive Hangar One, Naval Air Station Sunnyvale (Moffett Field), 1933. The military has long been a key presence and important source of patronage in the Santa Clara Valley. Courtesy of the Sunnyvale Historical Society and Museum Association.

Another crucial and often overlooked aspect of Silicon Valley's development was the influence of government—especially military—patronage. The military built several bases and coastal defense forts in the area, and during World War II the Bay Area became a huge shipbuilding center and port of embarkation for the Pacific theater (Figure 3).[6] Meanwhile, on the strength of World War II and Cold War military contracts, the valley's pioneering electronics firms grew from fledgling start-ups into serious players that began to challenge the supremacy of East Coast stalwarts like General Electric, RCA, and Westinghouse. Both Litton and Varian grew rapidly during the 1950s as a result of sales of their magnetron and klystron microwave tubes, which were used in military applications including radar, electronic countermeasures, and guided missile defense. Then, in 1956, Lockheed Missiles and Space opened up a manufacturing facility adjacent to the Moffett Field Naval Air Station in Sunnyvale to build advanced avionics for guided missiles and surveillance satellites. A new research and development (R&D) center near Stanford soon followed, and Lockheed rapidly became (by an order of magnitude) the largest employer in the valley, with 25,000 on the payroll by 1964.[7]

Indeed, as historian Stuart Leslie has argued, the Department of Defense was the valley's original "angel" investor and an ideal client. At a time when a six-figure venture capital investment was still considered fairly risky, a start-up like Varian could routinely win a $1 million contract from the Air Force or Navy, which paid top dollar for mission-critical defense technologies. Since the military was always looking for the latest cutting-edge technologies, these contracts typically included monies devoted specifically to R&D and further product development, not just production. Thus, the military essentially subsidized most technological innovation in the valley's burgeoning electronics industry, just at the federal armories had subsidized the advancement of precision manufacturing techniques through its contracts with Hartford arms maker Samuel Colt.[8]

Under the leadership of Palo Alto native Frederick Terman (1900–1982), Stanford also took maximum advantage of military and government patronage to become one of the nation's leading universities. After earning his Ph.D. at the Massachusetts Institute of Technology (MIT), Terman returned home and taught electrical engineering at Stanford, where he mentored entrepreneurs like Hewlett, Packard, Litton, and Varian. Later, as an ambitious dean and university provost, Terman rebuilt Stanford in MIT's image by aggressively pursuing federal research funds and cultivating close ties with local industry (Figure 4). For example, Terman recognized that Stanford could not match powerful rivals like Harvard and MIT in every science and engineering field, so he endeavored to create "steeples of excellence" in a few specific areas (such as electronics and aerospace) that would attract military grants. Terman encouraged his faculty to consult for local firms, arranged for local industrial researchers to teach specialized courses on campus, and set up an Honors Cooperative Program so that corporate employees could earn their degrees while working full time.[9]

FIGURE 4
Hewlett-Packard Company founders David "Dave" Packard (left) and William R. "Bill" Hewlett (center) with their Stanford University School of Engineering professor and mentor Fred Terman in 1952 at the door of the Hewlett-Packard wing of the Electronics Research Laboratory on the Stanford campus. Photo courtesy of Hewlett-Packard Company Archives.

Terman also helped establish two key institutions to encourage high-tech academic and corporate research. In 1946, Terman convinced the university trustees to launch the Stanford Research Institute (SRI), which performed contract research for government agencies, the various military branches, and industrial clients. Then in 1952, Terman opened the Stanford Industrial Park, a high-technology park on university land adjacent to campus. The park's earliest tenants included local firms looking to expand, including Hewlett-Packard and Varian Associates, as well as established East Coast firms such as Kodak, General Electric, and Sylvania attracted by the proximity to Stanford and the region's cutting-edge electronics scene. Altogether, through these intertwined relationships, Terman simultaneously built up Stanford and the valley's local electronics and aerospace industries through the "military-industrial-academic complex."[10] As discussed later in this book, academic entrepreneurs such as the University of Minnesota's C. Walton Lillehei (Medical Alley, Minnesota) and Colorado State University's Bryan Willson (Fort Collins, Colorado) have successfully pursued the same model of university-industrial cooperation that worked so well for Terman at Stanford.[11]

By the late 1950s and early 1960s, Stanford University and the region's pioneering electronics industry had helped create a critical mass of talented people, strong institutions, and ready resources in the Santa Clara Valley. Continuing advances in semiconductor technologies and their application to personal computing would put Silicon Valley on the map during the 1970s and 1980s.

PEOPLE
Academic Researchers, Corporate Scientists, and Amateur Hobbyists

The personal computer emerged from several Silicon Valley subcommunities. Stanford's academic researchers conceived of the mouse and graphical user interface that Xerox's corporate scientists built into the experimental Alto workstation. Meanwhile, engineers at Fairchild Semiconductor and Intel developed the mass-produced integrated circuits and microprocessors that powered the first inexpensive hobbyist kits that emerged from the Homebrew Computer Club. Apple's Macintosh computer integrated all of these influences, offering sophisticated features at an affordable price. Thus, a variety of inventors, including academic scientists, corporate researchers, and self-trained hobbyists, contributed to the invention of the personal computer in Silicon Valley.

So how *did* the Valley of Heart's Delight become Silicon Valley? No one is sure who originally coined the phrase, but the moniker Silicon Valley was certainly in use during the 1960s among East Coast engineers and defense contractors who traveled frequently to the Bay Area. In 1971, *Electronic News* reporter Don Hoefler wrote a three-part series on the history of the region's growing semiconductor industry, entitled "Silicon Valley U.S.A." Hoefler's article was the first appearance of the phrase in print, and the name stuck (Figure 5).[12]

Hoefler described how Palo Alto native William Shockley (1910–1989) (Figure 6) left his job at AT&T's Bell Labs and moved from the East Coast to establish Shockley Semiconductor in Mountain

> Government-sponsored research presents Stanford, and our School of Engineering, with a wonderful opportunity if we are prepared to exploit it.
>
> STANFORD UNIVERSITY'S
> FREDERICK TERMAN

One of the Fairchild Business Newspapers

Electronic News

Vol. 18 WHOLE No. 800   ★ ★ ★   New York, N.Y., Monday, January 11, 1970   TWENTY CENTS   One Year $5 Payable in Advance

C-5A Subs: The Waters Get Muddier

By JIM STROTHMAN

BURBANK, Calif. — Lockheed Aircraft Corp.'s decision last week to accept the Pentagon's Cheyenne Helicopter offer while rejecting the $200 million fixed-loss proposal for the C-5A has muddied the waters swirling around its major subs.

There was a flurry of activity in the strategy rooms of the subcontractors, but their deliberations were shrouded in silence to the outside world.

While Lockheed said it will go to the courts to seek a solution to its financial dispute with the government on the C-5A, some subs are wondering what Deputy Defense Secretary David Packard had in mind last month when he said he was seeking a solution which would not ruin Lockheed's subs, particularly on the C-5A.

A Lockheed spokesman said its C-5A electronics sub-

See C-5A, Page 12

SILICON VALLEY U.S.A.

(This is the first of a three-part series on the history of the semiconductor industry in the Bay Area, a behind-the-scenes report of the men, money and litigation which spawned 23 companies — from the fledgling rebels of Shockley Transistor to the present day.)

By DON C. HOEFLER

It was not a vintage year for semiconductor start-ups. Yet the 1970 year-end box score on the San Francisco Peninsula and Santa Clara Valley of California found four more new entries in the IC sweeps, one more than in 1969.

The pace has been so frantic that even hardened veterans of the semiconductor wars find it hard to realize that the Bay Area story covers an era of only 15 years. And only 23 years have passed since the invention of the transistor, which made it all possible.

For the story really begins on the day before Christmas Eve, Dec. 23, 1947. That was the day, at Bell Telephone Laboratories in Murray Hill, N.J., three distinguished scientists, Dr. John Bardeen, Dr. Walter Brattain and Dr. William Shockley, demonstrated the first successful transistor. It was made of germanium, a point-contact device that looked something like a crystal detector, complete with cat's whiskers.

The three inventors won the Nobel Prize for their efforts, but only one of them, Dr. Shockley, was determined to capitalize on the transistor commercially. In him lies the genesis of the San Francisco silicon story.

It was only by a quirk of fate, however, coupled with lack of management foresight, that Boston failed to become the major semiconductor center San Francisco is today. When Dr. Shockley left Bell Labs in 1954, he headed first for New England to become a consultant to Raytheon Co., with a view toward establishing a semiconductor firm there under its auspices.

His financial plan called for a guarantee to him of $1

Peripherals Firms Dig in Against IBM On Tape, Disk Price

Independent peripherals makers dug in last week for a long and strenuous price war with industry leader IBM.

The battlefield is the market for plug-compatible tape and disk drives, and the independents had to move their trenches back after a series of IBM advances involving cheaper models of both.

The list of companies that cut rental prices last week reads like a who's who of the independents —Telex, Potter, CalComp, Marshall and Tracor for openers, with Ampex and Memorex understood to be poised on the brink of cuts of their own.

This new tactic follows an earlier attack through legal channels when the independents' trade group, the Computer Peripheral Manufacturers Association, asked the Justice Department and Congress to take anti-trust action against IBM (EN, Dec. 28).

The targets are IBM's 2319 disk drive and 3420 tape drive families, which have left the independents with an exposed flank—namely prices higher than IBM's.

The price cuts among independents that came to light last week included a 22 per cent slash by Telex on its 5312 disks, to 9 per cent below IBM, and some 25 per cent on its IBM compatible tape drives, effective May 1.

Potter Instrument Co. immediately let it be known that it also was cutting prices and, furthermore, that some cuts would be effective at once and would amount to 10 per cent below IBM.

California Computer Products then got into the act with immediate 10 per cent cuts on its CD 14-12 disk in yet another assault on the IBM 2319.

FIGURE 5

In 1971, the region received its nickname when journalist Don Hoefler described the emergence of the semiconductor industry in a series of articles titled "Silicon Valley U.S.A." Source: Don C. Hoefler, "Silicon Valley U.S.A.," part 1, *Electronic News,* 11 January 1971, 1. © Don C. Hoefler. Photograph by Matthew Hersch.

FIGURE 6
Palo Alto native William Shockley, 1965. Shockley won the 1956 Nobel Prize in Physics for inventing the solid-state transistor with his AT&T colleagues and then established the first semiconductor firm in Silicon Valley. Chuck Painter/ Stanford News Service.

View. In 1947, Shockley (with John Bardeen and Walter Brattain) invented the solid-state transistor by taking advantage of the properties of semiconductors, materials like silicon and germanium whose ability to conduct (or not conduct) electrical current could be controlled by doping them with certain impurities and manipulating them with an electrical field. Like a vacuum tube, the transistor could send, receive, switch, and amplify electrical signals at various frequencies, but it was much smaller and more robust. As such, transistors were quickly adopted during the 1950s to help miniaturize all kinds of electronic appliances, including radios, hearing aids, and a key invention driving the development of Medical Alley, Minnesota—Earl Bakken's external, wearable pacemaker.[13]

Shockley might have located his start-up anywhere, but his mother, Stanford graduate Mary Bradford Shockley, still lived in the area; in addition, Stanford provost Frederick Terman actively recruited Shockley to locate near the university. Thus, in 1955, Shockley Semiconductor became the first semiconductor firm in the region. To staff his firm, Shockley recruited an immensely talented group of young and ambitious physicists and engineers with Ph.D.s from top graduate programs (MIT and University of California, Berkeley) and prior experience at key firms (Philco and Western Electric). Shockley was brilliant—he shared the Nobel Prize for Physics in 1956 for inventing the transistor—but his management style was abrasive. Thus, in 1957, eight of Shockley's young engineers resigned to start their own semiconductor firm. With initial backing from Boston investment bankers Bud Coyle and Arthur Rock, the so-called Traitorous Eight—Julius Blank, Victor Grinich, Jean Hoerni, Eugene Kleiner, Jay Last, Gordon Moore, Robert Noyce, and Sheldon Roberts—struck a deal with Fairchild Camera and Instrument to form a new start-up, Fairchild Semiconductor (Figure 7).[14]

The leader of this group was Robert Noyce. Noyce (1927–90) was born and raised in Iowa, the son of a minister. Noyce was a champion diver at Grinnell College then earned a Ph.D. in physics from MIT; for his brilliant mind (and effortless charm with the ladies) his graduate school friends nicknamed him Rapid Robert. In 1959, while at Fairchild Semiconductor, Noyce invented the first practical integrated circuit, or microchip—the electronic backbone of all modern computers and digital devices, from smartphones to digital watches.[15]

Meanwhile, as the semiconductor industry boomed in the 1960s, Fairchild Semiconductor emerged as the taproot of what Don Hoefler described as a "silicon tree" of electronics firms. For example, three of the original Shockley defectors—Jay Last, Jean Hoerni, and Sheldon Roberts—left Fairchild in 1961 to form Amelco; other "Fairchildren" would go on to found key semiconductor firms such as Advanced Micro Devices, National Semiconductor, and Intersil.[16] This spin-off phenomenon, so closely associated with Silicon Valley, is actually a common feature of many places of invention, in which ambitious inventors and entrepreneurs train first at a leading firm before leaving to found their own start-ups. As discussed later in this book, a similar process occurred among former machinists at Samuel Colt's armory in Hartford, Connecticut, and among former Medtronic employees in Medical Alley, Minnesota.

FIGURE 7

The so-called Traitorous Eight who left Shockley Semiconductor to found Fairchild Semiconductor, 1960. Robert Noyce sits front and center with his arm draped over his chair. Seated clockwise from Noyce are Jean Hoerni, Julius Blank, Victor Grinich, Eugene Kleiner, Gordon Moore, C. Sheldon Roberts, and Jay Last. © Wayne Miller/ Magnum Photos.

By far, the most famous of the Fairchild spin-offs was Intel, founded by Robert Noyce and Gordon Moore in 1968 (Figure 8). Noyce and Moore raised $2.5 million in financing by again working with Arthur Rock, who had moved to San Francisco to become the first high-tech venture capitalist on the West Coast.[17] After chafing under Shockley's domineering leadership, Noyce and his colleagues rejected the hierarchy of the established East Coast firms and embraced the egalitarian West Coast work culture already in place at pioneering firms like Hewlett-Packard, Litton Industries, and Varian Associates.

First, the dress code was casual: jeans, khakis, and open collars versus the East Coast suit and tie; later, software programmers in the 1980s and 1990s would push the boundaries of office informality by wearing T-shirts and sandals (Figure 9). There were no "sirs" or "misters" as workers and managers alike used first names across the board. In addition, local firms eschewed the East Coast's assigned parking spaces, closed-door offices, and executive dining rooms for open cubicles and a cafeteria where hourly rate fabrication workers and executives alike might share a table. Workers also enjoyed a high degree of autonomy; managers recruited talented and highly motivated engineers then got out of their way. Accordingly, employees—not just executives—enjoyed perks such as flexible working hours, generous health insurance and retirement benefits, stock options, profit sharing, and intramural corporate sports teams.[18]

FIGURE 8

Intel's executive team stands over a Rubylith of an integrated circuit, 1978. From left to right: Andy Grove, Robert Noyce, and Gordon Moore. Courtesy of Intel Corporation.

FIGURE 9

West Coast Computer Faire T-shirt, 1977. In the casual business culture of Silicon Valley, hobbyists and corporate programmers alike often wore T-shirts that expressed pride in various products, firms, and organizations. © 2014 Smithsonian Institution; photo by Richard Strauss. T-shirt design by Jim Warren, Eric Bakalinsky, and Bob Reiling. Courtesy of the Computer History Museum, Mountain View, California.

This informal work culture facilitated continuous innovation by encouraging the free flow of technical know-how within and among the valley's electronics firms. Inside the big East Coast firms, information flowed vertically through the hierarchy: reports up from workers and directives down from management, often with little exchange among the various divisions. In contrast, the valley's flat, egalitarian organizations typically eschewed formal reporting structures in favor of "management by walking around," in which a firm's most important communications occurred informally via "the ad hoc meetings that occur when we walk around the plant."[19] Information also flowed liberally between the valley's electronics firms. The booming electronics industry encouraged plenty of spin-offs and new start-up companies, resulting in extremely high job mobility, which, in turn, facilitated the rapid cross-pollination of technical knowledge across the region. When engineers moved between companies (or formed new ones), they brought the knowledge, skills, and experience they had acquired at their previous jobs.[20]

These exchanges were further encouraged by the valley's geography, which is hemmed in by the San Francisco Bay to the north and mountains on the east, south, and west. The resulting dense pattern of development decreased the physical distances between firms and increased the chance of serendipitous, informal contacts with other engineers and industry types—at parties, at church, at a kid's T-ball game.[21] But these exchanges occurred most frequently at the area's local watering holes. In a 1983 *Esquire* feature on Robert Noyce, Tom Wolfe wrote that "every year there was some place, the Wagon Wheel, Chez Yvonne, Rickey's, the Roundhouse, where members of this esoteric fraternity, the young men and women of the semiconductor industry, would head after work to have a drink and gossip and brag and trade war stories about phase jitters, phantom circuits, bubble memories," and other mysteries of the trade (Figure 10). Since techies changed jobs all the time and most shared a common Shockley/Fairchild ancestry, they were often more loyal to friends and former colleagues than their present employer. Thus, useful information flowed back and forth liberally, even among competitors.[22]

> We laugh about how often people change jobs. The joke is that you can change jobs and not change parking lots.
>
> SEMICONDUCTOR INDUSTRY VETERAN JEFFREY KALB

Meanwhile, outside the semiconductor industry, both academic and corporate researchers alike were taking advantage of the increased power and availability of integrated circuits to transform the nature of computing. For example, Douglas Engelbart established a research program at SRI to develop improvements in "human computer interaction." Engelbart (1925–2013) served as a Navy radar technician during World War II and gained his first exposure to reading and manipulating symbols on a screen. At SRI, Engelbart used grants from the Department of Defense's Advanced Research Projects Agency (ARPA) to extend this idea by inventing new input-output modes to replace punched cards and text-based command-line terminals. For example, Engelbart and his

FIGURE 10
Walker's Wagon Wheel bar in 1986. © 1986 by Carolyn Caddes, all rights reserved. Courtesy of the Department of Special Collections and University Archives, Stanford University Libraries.

SRI colleagues developed an early version of the graphical user interface (GUI), in which users manipulated pictures and symbols in an on-screen "window," and the "mouse," a now ubiquitous pointing and selection tool (Figure 11).[23]

The mouse and Engelbart's GUI concepts were eventually adopted commercially, first at Xerox, then later at Apple. Xerox, an East Coast copier company, established its Palo Alto Research Center (PARC) in 1970 to develop a new suite of computer and information technologies to supply "the office of the future." Xerox set about hiring top talent, including Bill English from Engelbart's Augmentation Research Laboratory and Alan Kay (b. 1940) (Figure 12), a new Ph.D. from the University of Utah's cutting-edge, ARPA-funded computer science program. In December 1972, Kay told *Rolling Stone* reporter (and former SRI associate) Stewart Brand, that his PARC colleagues were "really a frightening group, by far the best I know of as far as talent and creativity. The people here all have track records and are used to dealing lightning with both hands."[24]

In the 1970s, most mainframe computers were big, expensive, and shared by multiple users, typically elite technicians working at corporations and universities. However, as Brand reported "the general bent of research at Xerox" was "away from hugeness and centrality, toward the small and the personal, toward putting maximum computer power in the hands of every individual who wants it."[25] Accordingly, in 1973 Kay and his colleagues introduced the Xerox Alto, a "personal computer" that was small enough to fit on a desktop and incorporated improved versions of Engelbart's mouse and GUI (Figure 13). Moreover, Xerox connected several Altos via its "Ethernet," a local area network that enabled interoffice e-mail, file sharing, and network printing on PARC-developed laser printers.[26] Incredibly, Xerox never commercialized the Alto; in fact, several PARC researchers left Xerox to develop key technologies at new start-ups like 3Com (Ethernet) and Adobe (PostScript printing protocol). Although some critics have suggested that Xerox "fumbled the future," there is no doubting PARC's crucial influence on the development of personal computing.[27]

Alongside the highly educated and well-funded researchers at SRI and Xerox PARC, amateur hobbyists and "hackers" were also influential in the invention of the personal computer. Silicon Valley's hobbyist community was deeply influenced by the hippie counterculture that emerged in and around San Francisco in the 1960s and 1970s. But unlike the hippies who retreated to the communes and rejected technology, the hobbyists loved computers but lamented that their high cost (e.g., $6,000 for a Digital PDP-8 minicomputer) typically confined their use to the military, big insurance

FIGURE 12
Inventor Alan Kay sits in his office at
Xerox's Palo Alto Research Center (PARC)
in 1973. Kay was a driving force behind the
Xerox Alto personal computer, shown at
left. Courtesy of PARC, a Xerox company.

FIGURE 13
The Xerox Alto, 1973. © 2014 Smithsonian Institution;
photo by Jaclyn Nash. Courtesy of National
Museum of American History.

companies, and other establishment institutions.[28] Some groups pooled their resources, acquired a computer, and operated like a co-op. For example, Bob Albrecht, leader of the People's Computer Company, procured a PDP-8, set everything up in a Menlo Park storefront, added some bean bags and a few bookshelves with technical literature, and invited hackers, students, and housewives alike to come inside and try computing for themselves. Other groups, like the Homebrew Computer Club, formed around the idea that hobbyists could build their own computers from scratch.[29]

The Homebrew Computer Club formed shortly after the appearance of the MITS Altair 8800 computer kit on the January 1975 cover of *Popular Electronics*, which called it the "world's first minicomputer kit to rival commercial models" (Figure 14). The Altair 8800 kit was built around Intel's 8080 microprocessor by Micro Instrumentation Telemetry Systems (MITS) of Albuquerque and is often described as the first personal computer. At $397 ($498 fully assembled) the Altair kit was affordable but it was difficult to assemble and had limited (even primitive) capabilities. Nevertheless, for many hobbyists, the Altair 8800 realized the dream of a computer cheap enough for anyone to own and control for themselves (Figure 15).[30]

Thus, in the spring of 1975, just weeks after *Popular Electronics* announced the Altair 8800, hobbyists Fred Moore and Gordon French tacked up an announcement on corkboards all around Silicon Valley. It read,

> AMATEUR COMPUTER USERS GROUP
> HOMEBREW COMPUTER CLUB . . . you name it.
> Are you building your own computer? Terminal? TV Typewriter? I/O Device? Or some other digital black magic box? Or are you buying time on a time-sharing service?
> If so, you might like to come to a gathering of people with likeminded interests.
> Exchange information, swap ideas, help work on a project, whatever . . .

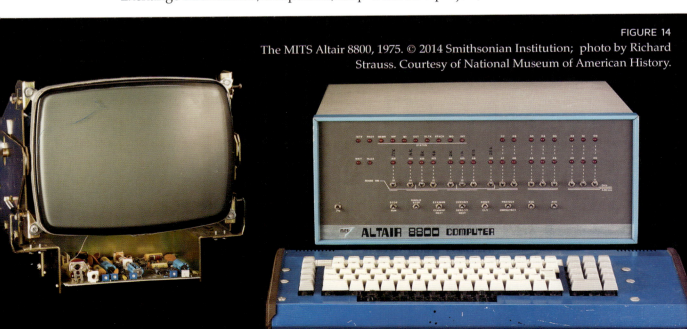

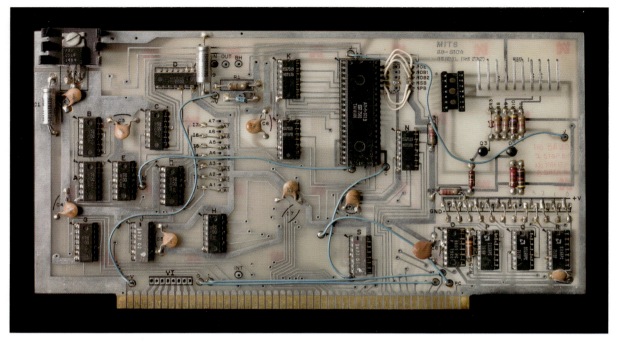

MITS Altair 8800 circuit board, 1975. The Altair's open architecture encouraged experimentation. Hobbyists expanded the kit's capabilities by designing custom-built circuit boards to add a keyboard interface or extra memory. © 2014 Smithsonian Institution; photo by Richard Strauss. Courtesy of National Museum of American History.

The first meeting of the Homebrew Computer Club convened on the evening of 5 March 1975; all 32 attendees fit easily into French's two-car garage in Menlo Park. The Homebrew Computer Club decided to meet every two weeks, and it got bigger and bigger, eventually moving to various auditoriums at Stanford University to accommodate more than 100 attendees.[31]

The group was open to everyone and was somewhat anarchic—there were no dues, no by-laws, and no official membership. The club's free newsletter, edited by Moore, became a crucial source of information on microcomputers and a link that connected the region's hobbyists (Figure 16). Lee Felsenstein eventually took over for French as master of ceremonies and presided over the meetings. First came a "mapping" session, in which individuals introduced themselves and asked questions or offered something to sell, trade, or give away. This was followed by a formal presentation, usually of someone's latest invention. Finally, there was a "random access" session where the attendees scrambled around the auditorium to meet others with common interests. After the meetings, several members would often continue the conversations at the Oasis, a burgers and beer joint still located near the Stanford campus at 241 El Camino Real in Menlo Park.[32]

In short, the Homebrew Computer Club became a remarkable arena for exchanging and sharing information about early personal computers. As member Steve Wozniak remembered, "The

AMATEUR COMPUTER USERS GROUP   NEWSLETTER   HOMEBREW COMPUTER CLUB
Issue number two   Fred Moore, editor,   558 Santa Cruz Ave., Menlo Park, Ca. 94025   April 12, 1975

FIGURE 16
Cover of the *Homebrew Computer Club Newsletter*, 15 April 1975. © Lee Felsenstein. Courtesy of the Liza Loop Papers, Collection M1141, Department of Special Collections and University Archives, Stanford University Libraries.

theme of the club was 'Give to help others.' During the 'random access period' ... you would wander outside and find people trading devices or information and helping each other."[33] Several Silicon Valley start-ups were born from this potent group. For example, Stanford graduates Harry Garland and Roger Melen founded Cromemco in 1974 to build add-ons for the Altair, such as their Cyclops camera and Dazzler graphics card. Likewise, Lee Felsenstein, Gordon French, Bob Marsh, and Steve Dompier started Processor Technology Corporation and built the Sol-20 microcomputer to compete with the Altair. Altogether, more than twenty firms originated from the membership of the Homebrew Computer Club. Of course, the most famous and enduring Homebrew start-up was Apple Computer.[34]

Apple was founded by Steve Wozniak (b. 1950) and fellow Homebrew Club member, Steve Jobs (1955–2011) (Figure 17). The two Steves met as teenagers and bonded quickly over their mutual interest in electronics. Both teenagers bounced in and out of college: Wozniak attended the

University of Colorado and De Anza College before dropping out of Berkeley; Jobs spent some time at Reed College before dropping out to spend a year in India. By the early 1970s, both young men were back working in the Bay Area, Wozniak at Hewlett-Packard and Jobs at Atari, the video game company.[35]

Like other Silicon Valley hobbyists, Wozniak and Jobs were extremely excited about the appearance of the Altair 8800 in 1975. However, even at $400 the price tag was too steep for Wozniak, so he decided to build his own kit. Wozniak could not afford an Intel 8080 chip set, so he built his kit around the cheaper MOS Technologies 6502 microprocessor, which he bought at a trade show for $20. Wozniak and Jobs "liberated" some other spare parts from HP and Atari, and when the prototype kit was completed, Wozniak debuted the Apple I at the Homebrew Computer Club in February 1976. Like other hobbyist kits, the Apple I was very simple, a stripped down circuit board with 4K of RAM; users would have to supply their own power supply, keyboard, storage system, and the BASIC programming language via cassette tape interface. Sensing an opportunity, Jobs made a successful sales pitch to Paul Terrell, owner of the Byte Shop retail computer store in Mountain

FIGURE 17
Apple Computer founders Steve Jobs (left) and Steve Wozniak (right), 1976. © DB Apple/dpa/Corbis.

View. Terrell had witnessed the Homebrew demonstration and agreed to buy 50 Apple I kits at $500 apiece, with delivery due in 30 days. Jobs sold his Volkswagen van, Wozniak sold his HP calculator, and they cobbled together $1,300 in start-up capital. Wozniak and Jobs retreated to the Jobs family garage in Los Altos, assembled the kits, delivered them to Terrell in 29 days, collected $25,000, paid off their parts supplier, and kept a modest profit. Jobs was 21 years old; Wozniak was 25.[36]

Meanwhile, Wozniak continued to improve his designs, and Apple Computer grew rapidly. The Apple II, introduced in 1977, became a big seller, and as the firm's annual sales surpassed $70 million in 1979, Apple prepared to take the company public.[37] In 1979, Xerox executives approached Apple and asked to make a $1million investment in the company before the initial public offering, anticipating big returns. Jobs agreed on the condition that Xerox "open its kimono"—he wanted a tour of the PARC skunk works. Over the objections of several PARC engineers, Jobs and a few Apple lieutenants were given demonstrations of the Xerox Alto and its successor the Xerox Star on two different visits in December 1979. Years later, Jobs reminisced about those tours and called the Alto's GUI "the best thing I'd ever seen in my life. . . . [Within] ten minutes it was obvious to me that all computers would work like this someday."[38]

> I met Woz when I was maybe 12 years old, 13 years old. He was the first person I met who knew more about electronics than I did, so we became fast friends.
>
> APPLE COFOUNDER
> STEVE JOBS

It would be convenient to draw a straight line from Jobs's Xerox PARC visit to the Apple Macintosh, but the details of the Mac's lineage are far more complex. First, Apple had already been experimenting with the GUI for its next model, the Apple Lisa. The Lisa, eventually introduced in 1983, featured a mouse and GUI but retailed for the shockingly high price of $12,000, ten times the price of an Apple II. Plus, the Apple Macintosh project, originally led by Jef Raskin, was already underway in 1979 by the time Jobs toured PARC. Raskin, a former computer science professor and PARC researcher, had performed research on GUIs and had been lobbying Apple executives to build a low-cost, easy-to-use computer for the masses that would retail for under $1,500. Raskin had assembled a talented and cohesive team of developers, including Bill Atkinson, Andy Hertzfeld, Joanna Hoffman, Brian Howard, Burrell Smith, and Bud Tribble, and work was already underway by 1981, when Jobs forced out Raskin and took control of the Macintosh project.[39] Thus, one might characterize Xerox as naive and Jobs as a pirate, but it is more accurate to suggest that Jobs's PARC demonstration validated and kick-started ideas already brewing at Apple.

By this time, Wozniak was largely out of the picture and recovering from amnesia after crashing his private plane in February 1981. Meanwhile, Jobs inherited the Macintosh team and drove them hard to create an "insanely great" masterpiece that would combine the power of Xerox's workstations at Raskin's cheaper price point.[40] Meanwhile, graphic designer Susan Kare joined the team in 1983 and designed the distinctive icons, typefaces, and other graphic elements that made the Macintosh so intuitive and user-friendly (Figure 18). Like the Xerox Alto, the Apple Macintosh featured a bit-mapped display in which each pixel on the screen was individually controlled by a single bit of computer data. Creating graphics was simply a matter of deciding which bits to turn on and off. Kare used 32×32 blocks of graph paper (and later the Mac's icon editor) to design pictorial metaphors for various commands and operating statuses; these became the Mac's signature icons, such as the trash can, the paintbrush, and the ticking bomb (Figure 19). She also designed several proportional typefaces (e.g., Geneva, New York, and Monaco) that improved upon the monospaced characters found on typewriters and earlier computers.[41]

The Apple Macintosh was introduced in January 1984 with great fanfare, including a block-buster television ad during the Super Bowl inspired by *1984,* George Orwell's dystopian novel. At the reasonably affordable price of $2,495, the Macintosh *commercialized* several revolutionary features we now take for granted: a handheld input device called a mouse; a GUI with overlapping windows, menus, and cut-copy-paste editing; and Kare's clickable icons and multiple typefaces (Figure 20). Moreover, the Macintosh represented the synthesis of several Silicon Valley subcommunities—the academic explorations of Douglas Engelbart's team at SRI, the corporate research by Alan Kay and others at Xerox PARC, and the do-it-yourself experimentation of the Homebrew Club's hobbyists—who together helped shape the personal computer.[42]

FIGURE 18
Susan Kare at her desk at Apple Computer, Cupertino, California, 1984. Kare designed most of the icons and fonts for the Apple Macintosh (two of which sit on the desk behind her). © Norman Seeff.

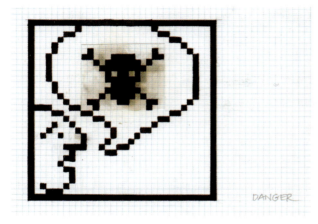

FIGURE 19
An original pencil sketch for a "Danger" icon from
Susan Kare's graph paper sketchbook, about 1983.
Courtesy of Susan Kare and kareprints.com.

FIGURE 20
Apple Macintosh computer, 1984. © 2014 Smithsonian
Institution; photo by Richard Strauss. Courtesy of
National Museum of American History.

INVENTION

The Personal Computer

During the 1960s and 1970s, mainframe computers were big and expensive, and only the government and big businesses could afford them. As such, only an elite, highly trained priesthood of
technicians and programmers tended to the mysterious machines, typing cryptic program codes
into terminals and gathering the resulting output on punched cards and paper tape. Today, personal computers are smaller, more affordable, and much easier to use, thanks to inventions developed in Silicon Valley. First, the invention of transistors, followed by integrated circuits and
microprocessors, provided the electronic infrastructure for the digital age. Second, continuing advances in these microchips made computers personal as the hardware became smaller, faster, and
cheaper. Third, advances in human-computer interaction such as the handheld mouse and GUI
made personal computers less intimidating and more user-friendly.

The personal computer revolution arguably began with the invention of the transistor at Bell
Labs in 1947. Transistors replaced the bulky vacuum tubes used for memory and processing and
helped shrink massive, room-sized computers down to personal scale (Figure 21). The transistor,
invented by William Shockley, Walter Brattain, and John Bardeen, could either conduct—or not
conduct—electricity in a given direction, essentially turning the device into a switch (1 = on, 0
= off). Using the binary numbering system, a sequence of transistors could represent numbers,
characters, or any kind of digital data (e.g., lowercase a = 01100001). If one assembled enough

transistors in a circuit, a digital computer could store text and numbers in memory and run programs to manipulate them.

In the first computer circuits, the individual transistors, diodes, and resistors were attached to each other, one at a time, by hand in a process fraught with errors and failures. In contrast, Robert Noyce's *integrated* circuit was a complete electronic circuit printed in metal directly on the flat, two-dimensional plane of a small silicon wafer using photolithography techniques. In this so-called planar process, the circuit's individual components could be printed and connected to each other simultaneously. Thus, integrated circuits were much more reliable than standard hand-fabricated circuits, plus they could be much smaller than the scale of connections achievable by human hands (Figure 22).[43]

As Noyce and his Fairchild colleagues refined and improved the planar manufacturing process, they were able to fit more and more components onto the same fingernail-sized chips. Fairchild's R&D chief, Gordon Moore, described this phenomenon in a 1965 article for *Electronics* magazine entitled "Cramming More Components onto Integrated Circuits." Moore observed that since Noyce invented the integrated circuit in 1959, the number of components that could be placed on an integrated circuit had doubled approximately every two years; furthermore, he expected this pace to continue indefinitely.[44] The implication of "Moore's law" was that microchips would always be getting smaller, faster, cheaper, and more powerful and would do so in a *predictable* fashion. Thus, Moore's law was hugely important for new product design because inventors could spec out devices with certain performance parameters and be confident that the necessary microchips would be available in due time.[45]

The subsequent invention of the microprocessor at Intel was a crucial development that enabled the emergence of personal "microcomputers" by tipping the scales both technically and economically. In 1969, a Japanese firm named Busicom hired Intel to build the chip sets for a scientific calculator. Each of the calculator's trigonometric and square root functions would require its own new custom-designed chip. Instead, Intel designer Marcian "Ted" Hoff suggested a more elegant solution: why not design one *general-purpose* chip that could be programmed with the specific calculator functions? The resulting Intel 4004 chip set was not just an integrated circuit but a rudimentary programmable computer in its own right. Other semiconductor firms quickly adopted Intel's microprocessor concept. The resulting competition (and the onward march of Moore's law) drove down the price for processing power and paved the way for cheaper personal computers such as the Altair 8800 and Apple I kits built by Homebrew Club hobbyists (Figure 23).

The availability of enhanced processing power also enabled researchers such as SRI's Douglas Engelbart to invent improvements in human-computer interaction. For example, Engelbart and his team developed the so-called chord keyset, an efficient five-button keyboard that Engelbart believed was superior to the standard QWERTY keyboard. Then, to complement textual command-line instructions, Engelbart developed an early version of the GUI, in which users manipulated

WHAT IS A TRANSISTOR?

The small device attached below is a transistor of the type manufactured at this plant. Transistors belong to a family of electronic components called semiconductors. They are used in various electronic circuits to switch, amplify or otherwise control the flow of current. Transistors do many jobs formerly assigned to vacuum tubes.

The die is the functional "heart" of the transistor. Ours is made of silicon, a semiconductor material. We diffuse certain chemicals into the silicon to form two "junctions" in the die. These junctions (see diagram at left) are what make a semiconductor material a transistor. When power is applied to the leads the electrical interaction at the junctions causes transistor to function.

The transistors are carefully tested and classified prior to sale. We sell transistors to electronic equipment manufacturers for use in equipment such as computers.

DOUBLE DIFFUSED SILICON PLANAR STRUCTURE

DIFFUSED TRANSISTOR DIE (ENLARGED DIAGRAM)

WINDOWS   MASK

JUNCTIONS (CROSS SECTION)

METAL CONTACTS

(TOP VIEW)

GOLD LEADWIRES

DIE

GOLD SOLDER

CAP

HEADER

HEADER WITH DIE AND LEADWIRES ATTACHED

FINISHED PRODUCT

FAIRCHILD SEMICONDUCTOR CORPORATION
COPYRIGHT 1961

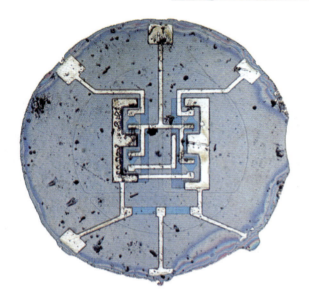

FIGURE 22

Fairchild Semiconductor's first commercial integrated circuit was a simple "flip-flop" switch, 1960. © Fairchild Semiconductor. Courtesy of the Computer History Museum, Mountain View, California.

pictures and symbols in an on-screen window. To manipulate the on-screen symbols and text, Engelbart and his assistant Bill English built a handheld pointing device in 1964. This prototype mouse was a simple wooden box with two perpendicular metal wheels, a selection button, and a wire connection to the processor (Figure 24).[46]

Alan Kay and his colleagues refined and extended Engelbart's ideas at Xerox PARC; in turn, Steve Jobs, Susan Kare, and their colleagues at Apple Computer commercialized them. Thus, the Xerox Alto (1973) and the Apple Macintosh (1984) incorporated nearly every major feature of personal computers we now take for granted: a desktop-sized computer with an individual user versus time-sharing on a closet-sized mainframe; a point-and-click mouse and QWERTY keyboard for input; a rich bit-mapped display (versus simple text, punch cards, or paper tape) for output; a GUI featuring overlapping windows, pull-down menus, symbolic icons, and multiple typefaces; cut-copy-paste editing and what you see is what you get (WYSIWYG) desktop publishing and printing. In addition, the experimental Alto (although not the commercial Macintosh) also featured PARC's

FIGURE 24
A replica of the first mouse prototype, invented by Douglas Engelbart and built by Bill English at the Stanford Research Institute (SRI) in 1964. © 2014 Smithsonian Institution; photo by Richard Strauss. Courtesy of National Museum of American History.

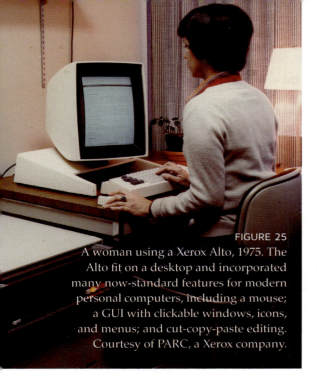

Ethernet hardware and packet-switching protocol, which enabled local area networks for interoffice e-mail, file sharing, and network connections to laser printers.

Altogether, during the crucial period between the early 1960s and the mid-1980s, academic researchers, corporate engineers, and amateur hobbyists in Silicon Valley developed new inventions in miniature digital circuitry, computer hardware, and interactive software that coalesced into the modern personal computer (Figure 25).

SILICON VALLEY TODAY
Continuous Innovation

Today, Silicon Valley remains one of the world's most vibrant entrepreneurial regions and a source of continuous high-tech innovation. It is still our best-known place of invention. Like many of the regions highlighted in this book, Silicon Valley benefited greatly from what economists call "path dependence" or "agglomeration effects." These concepts suggest that the success of an early technical or economic development snowballs from an initial advantage into a critical mass of wealth, talent, institutions, and know-how. Those assets are subsequently reinvested into the next generation of invention and wealth creation ad infinitum in a continuous virtuous cycle.[47]

We can see this clearly in Silicon Valley. The region's climate and geographical endowments made it an agricultural breadbasket. When gold was discovered, the railroads came, and that wealth built Stanford University. Stanford, via Fredrick Terman, encouraged its graduates to found several pioneering electronics firms, including Hewlett-Packard, Litton Industries, and Varian Associates. After World War II, Stanford and these firms were positioned to take advantage of postwar military and government investments in high-technology research. Terman, in turn, recruited Palo Alto native William Shockley to build his transistor business among the valley's burgeoning electronics industry. The talent that Shockley nurtured (and ran off) at Shockley Semiconductor included the founders of Fairchild Semiconductor, which, in turn, groomed the founders of Intel, National Semiconductor, and Advanced Micro Devices. Douglas Engelbart, an academic researcher at SRI, leveraged government grants and advances in semiconductor technology to create new ways of interfacing with computers. In turn, Engelbart's ideas were adopted and extended at Xerox PARC. With further advances in semiconductor technology (and the onward march of Moore's law), microprocessors became cheap enough that hobbyists at the Homebrew Computer Club could build their own personal computers. Finally, Apple's Macintosh synthesized both the

FIGURE 26

In the Chips board game, 1980. In this game, players passed through actual Silicon Valley electronics firms, banks, and landmarks, trying to "make their wealth" by successfully negotiating various opportunities and pitfalls.
© 2014 Smithsonian Institution; photo by Richard Strauss. Courtesy of National Museum of American History.

research and hobbyist cultures that emerged in Silicon Valley during the 1970s and 1980s and commercialized the paradigm of personal computing.

Thus, by the 1980s, Silicon Valley had rightfully earned its reputation as the leading high-technology region in the world (Figure 26). Similar to California's earlier Gold Rush, the region had become an economic juggernaut, where shaggy 25-year olds in T-shirts routinely became instant millionaires. However, the valley also had a darker side. With few exceptions (such as Apple's Joanna Hoffman and Susan Kare), the high-tech industry was a veritable boys' club, with huge gender imbalances among the ranks of highly skilled engineers and programmers. Meanwhile, as the valley's inventors cashed in, its low-skilled workers often struggled. Many of the processes used to fabricate electronics and computer components are highly toxic, which endangered factory workers and increasingly marred the former Valley of Heart's Delight with polluted air, land, and streams. And unlike the orchards and canneries they replaced, the high-tech industries were successful in preventing unionization, saddling these workers with low pay in a region with some of the nation's highest housing prices. But even these hazardous, low-paid jobs began to disappear as high-tech firms increasingly moved their production facilities offshore to Taiwan and mainland China.[48]

Despite these challenges, Silicon Valley has remained the epicenter of the high-tech economy. The region has capitalized on the technical advances, institutions, know-how, and wealth created from its past achievements to maintain and even accelerate its furious pace of innovation. In the 1980s, the rapid adoption of personal computers created a demand for better software applications, so the valley's coders developed everything from back-office accounting applications (Oracle) to video games (Electronic Arts). In the 1990s, the federal government lifted restrictions that limited use of the ARPANet, a government-funded computer communications network that was developed in part at SRI. As the Internet became commercially available, it spawned a whole generation of innovative Silicon Valley enterprises. These enterprises included new businesses in browsers (Netscape), search engines (Google, Yahoo!), e-commerce (eBay, PayPal), and social media (Facebook, Twitter, YouTube). The Internet also made it faster and cheaper to deliver all kinds of content, so Silicon Valley services such as Apple's iTunes and Netflix pioneered in the distribution and streaming of digital music and movies. Prior to his death in 2011, founder Steve Jobs spearheaded the reinvention of Apple, which dropped Computer from its corporate title and drew immense profits from its iPods, iPhones, and iPads, making the Homebrew start-up one of the world's most valuable companies. Indeed, when it comes to continuous innovation, Silicon Valley shows no signs of slowing down.[49]

> Ready or not, computers are coming to the people. That's good news, maybe the best since psychedelics.
>
> ROLLING STONE REPORTER AND COUNTERCULTURAL ICON STEWART BRAND

BRONX
Neighborhood Streets Create New Beats
NEW YORK (1970s)

Laurel Fritzsch

The Bronx in the 1970s was culturally rich with communities of predominantly African Americans, immigrants from Caribbean islands such as Puerto Rico and Jamaica, and historically Poles, Italians, and Russians. Sections were also economically devastated. This circumstance made the Bronx the right environment for the invention of hip-hop. The residents' diverse heritage influenced the sound of hip-hop while the urban landscape provided the raw materials for its technical innovations. Disc jockeys (DJs) used their families' sound systems or parts of them or mined the Bronx's abandoned buildings, cars, and streets for the components they needed to craft the "best" sound system. Inexpensive turntables, speakers, and components and scavenged materials were reworked, reimagined, and rewired in ways never seen before. The transformed equipment was paired with newly created techniques for manipulating records. The result was a system that made—not just played—a new kind of music.

DJs worked on their sound systems and techniques first in private spaces such as apartments, then made neighborhood spaces such as community rooms in public housing, parks, streets, and schools their stage. DJs Kool Herc, Afrika Bambaataa, Grandmaster Flash, and GrandWizzard Theodore each contributed to the development of some now iconic aspects of hip-hop deejaying. Some of their contributions include DJ Kool Herc's massive bass-heavy sound system; DJ Afrika Bambaataa's mixing selections of songs and sounds; DJ Grandmaster Flash's physical manipulation of records, quick mixing, and peek-a-boo system; and DJ GrandWizzard Theodore's "scratch." Their innovations created more than music—they defined what a DJ could do, and DJs became mentors, neighborhood celebrities, and sources of inspiration. Today, the sound systems, techniques, and genre they invented are global in scale and diversity.

PLACE

A Brief History of the Bronx

The Bronx is the northernmost of New York City's five boroughs and the only one primarily on the mainland. Separated from Manhattan Island by the Harlem River to the west and the East River to the south, the Bronx itself is segmented by hills and the Bronx River that runs down its middle. Although people have always lived in the Bronx, a large influx of primarily European immigrants settled there in the early 1900s. In the years following World War II, moderate- and high-income Bronxites left, and primarily African Americans and West Indies immigrants moved in. It took only a few years for poor city planning, the 1960s heroin epidemic, and other upheavals to radically alter the Bronx's working-class neighborhoods.[1]

Perhaps the most noticeable contribution to the borough's decline came in 1959 when city, state, and federal authorities began executing the construction of the Cross Bronx Expressway, which displaced Bronx residents, lowered property values around it, encouraged buildings to be abandoned, and further physically segmented and marginalized the already geographically isolated Bronx from the rest of New York City (Figure 1).[2] Additionally, the city's welfare department massed extreme

FIGURE 1
The 3rd Avenue El above the Cross Bronx Expressway, 1974. Photograph by Jack Boucher for the National Park Service. Courtesy Library of Congress, Prints and Photographs Division, HAER, Reproduction number HAER NY,3-BRONX,13-45.

poverty in the Bronx by placing impoverished newly arrived Puerto Rican and southern African American immigrants in south Bronx tenement buildings.[3] Landlords typically could not afford to maintain their buildings and sometimes burned them down for the insurance money. The Bronx's job market also declined.[4] As the Bronx fell into an economic depression, city officials hastened its demise through a policy called "planned shrinkage," which proposed to expedite the "death" and eventual redevelopment of these neighborhoods by removing services, including police, firefighting, and subway services, from "sick neighborhoods" (which happened to be home to primarily poor and nonwhite communities in the South Bronx, Harlem, and Brooklyn).[5]

As a result of these policies, the 1970s Bronx resembled a war zone with street gangs ruling the borough, vacant buildings, and abandoned cars whose owners could no longer afford to maintain them (Figure 2). However, there was a vital and supportive community of diverse people in the Bronx living their lives and calling it home:

> The vibe of the South Bronx at the time was very colorful, animated, happy-go-lucky, do what you want. I used to tell people I was from the South Bronx and they were like, "Oh, abandoned buildings, gangsters," but on the contrary— it was family back

FIGURE 2
East 140th Street at St. Ann's Avenue, the Bronx, 1977. Photograph © Camilo José Vergara.

then. Yes, there were gangs and drugs back then, but there are gangs and drugs everywhere in this country. We were just more out in the open. But it was a good time. There were family values back then. . . . Yes the Bronx was burning at the time. You didn't know if your building was going to be there the next day, because if you lived in an old tenement, your building might be marked for arson. But it was part of the time. You just picked up and moved on.[6] (Joe Conzo)

Being a child of the '60s and '70s, I grew up in the Bronx when the worst things you could say about the Bronx were true. Abandoned buildings and burnt-down remains. The greatest city in the world was bankrupt and corrupt. But from these conditions arose a culture that would eventually take the world by storm.[7] (GrandMaster Caz)

> Hip-hop came from nothing. The people that created hip-hop had nothing. And what they did was, they created something from nothing.
> GRANDWIZZARD THEODORE

PEOPLE

The community in the Bronx during the 1970s had largely African American, Latino, and Caribbean roots (approximately one-third African American and slightly more than one-third Latino in 1970), which contributed to the Bronx becoming the central location where the elements of hip-hop came together.[8] In hip-hop's infancy women were primarily b-girls, aerosol artists, or MCs, but very few were DJs. Aspects of hip-hop music can be attributed to multiple origins, but they combined to become what is culturally defined as hip-hop in the Bronx. Some of the most immediate influences to what was defined as hip-hop came from New York City night clubs, discos, and discotheques; Jamaican immigrants; and b-boys and b-girls (described by the media as break-dancers).

New York City DJs

New York City (NYC) DJs, who primarily played disco music during the 1970s, made a strong immediate contribution to hip-hop. Disco DJs were the first to experiment with using techniques in live performances that were already in use by radio DJs. For example, Brooklyn-born DJ Francis Grasso used the slip cueing technique, also known as beat matching (using a slip mat to hold a record in place before releasing it at the exact desired point to transition smoothly between two different songs), live in clubs.[9] Club DJs also used equipment with the latest technological features such as a pitch adjuster that allowed for a smoother transition between songs.[10] Although the goal of disco DJs' techniques was to keep the music flowing seamlessly, their techniques inspired hip-hop DJs who liked abrupt music transitions.[11] Musically, in disco, as in hip-hop, rhythm dominates over the melody.[12] Disco DJs motivated the record labels to make versions of the most popular songs with vocal and instrumental versions for club use only.[13] However, most Bronx teenagers were too young to get into the popular Disco clubs and resented their dress codes, motivating the teens to create an alternative. As DJ Disco Wiz recalls, "We weren't socially accepted at disco joints; we were pretty much segregated. I was looking for an outlet to express myself."[14] New York City youth and young adults decided to take their parties to parks and community rooms.

Jamaican Community and DJ Kool Herc

Immigrants from Jamaica who grew up in the Jamaican music scene had a significant influence on what became hip-hop music. DJ Kool Herc, who is considered the father of hip-hop, was born in Kingston, Jamaica, with the name Clive Campbell and immigrated to the United States as a child. One feature of Kingston music makers was their mobile sound systems with massive speakers built into wooden cabinets that were designed to play music outside (Figure 3).[15] The music played

FIGURE 3
Ken Davey with the Mutt and Jeff sound system, the Alpha Boys' School, Kingston, Jamaica, circa 1959. Courtesy of the EMP Museum, Seattle, Washington.

Places of Invention

by the sound system's operator was typically bass and drum heavy, which is best for sound traveling long distances.[16]

Although other DJs with large systems were in the Bronx prior to DJ Kool Herc, Jamaican systems were one inspiration for Kool Herc to build what became the largest and most formidable sound system in the Bronx.[17] DJ Kool Herc imitated the bass-heavy systems of his native country, creating one of the first distinguishing hip-hop characteristics: big sound systems with heavy bass.[18] With a borrowed Shure P.A. (a brand of public address system) and guitar amps, he built his own mixer using the guitar amp to switch from one turntable to the other. As Kool Herc explains, "What they was doing was using one of the channels from the turntables and using the brain itself to power the whole thing . . . I used the pre-amp, used the speakers wires put into a channel, and used the two knobs to mix. I got more sound than they ever got."[19]

> When Kool Herc finally hit the scene, we started getting the buzz that something was different. The funk that he threw on turntables, and the soul that came across with the African beats, was something that I related to. I could feel it.
>
> DJ DISCO WIZ

Kool Herc was the DJ at a fundraising party for his sister Cindy in the rec room of their apartment building at 1520 Sedgwick Avenue on 11 August 1973 that took on folklore status as the true beginning of hip-hop (Figure 4). Kool Herc spun records, and Cindy charged a few cents admission at the door.[20] As Kool Herc held more rec room parties and word of his sound spread, he outgrew the rec room (which also could not be rented very frequently) and took his sound system to the parks of the Bronx.[21]

Kool Herc's sound system soon became the envy and aspiration of other DJs. As DJ Disco Wiz explains, "For us, for the first-generation pioneers it was always about acquiring that sound system. That was the quest. The awesome amp, the awesome turntables, the right mixer, the right speakers

FIGURE 4
The 1520 Sedgwick Avenue Apartments, 2010. Bronx high-rise adjacent the Cross-Bronx Expressway where DJ Kool Herc and his family lived and where Herc played his first hip-hop parties in the building's recreational room. Photograph by Amanda Murray. © 2014 Smithsonian Institution; photo by Amanda Murray. Courtesy of National Museum of American History.

towering—you wanted to emulate what Kool Herc had."[22] The lack of affordable equipment required aspiring DJs to use a combination of secondhand, borrowed, or inexpensive equipment and scavenged materials to build their systems and make it sound impressive. Like the amateur hobbyists in Silicon Valley did with computers, some collaborating would-be DJs pooled their resources to build sound systems. For example, "when we first started, it was two bass speakers, raw horns, and a guitar amp. And it was loud. But Herc had a real system," recalls DJ Tony Tone from the Cold Crush Brothers (Figure 5). "So we made some money, and we went and bought equipment.

We found out about crossover power amps; we tried to get more professional. We had 15-inch speakers made out of 55-gallon steel drums. They laid on the floor and they had 6 inch legs that shot 'em up. The speakers faced down so you could get a bass reflex, and it would shoot out—you could hear it for at least ten blocks."[23] DJ Baron recalls that the speakers "made a lot of noise, but it was just the uniqueness of a garbage can as a speaker."[24] Similar to the Frankenstein computers in Silicon Valley, the result was DJs with unique custom-made sound systems.

Another aspect of Jamaican music makers that Kool Herc adopted was a practice they called "toasting." To elevate the crowd's excitement, toasters would shout phrases or make staccato noises over a song's instrumental break to highlight its rhythm.[25] DJ Kool Herc adopted this, but found that pulling double duty as both DJ and MC was difficult, and consequently, he employed an MC named Coke La Rock.[26] Although Coke La Rock emceed in a style less like modern rap and more like the Jamaican style of toasting, once the idea of reintroducing words over instrumental breaks started, the next set of MCs was inspired to first develop catchphrases, then chant rhymes, and eventually create song-length poems.[27] Kool Herc's sound system and employment of an MC gave hip-hop some of its defining characteristics that others would soon build on.

B-boys and B-girls

Latino immigrants' music influenced early hip-hop's sound in the Bronx. In particular, Puerto Rican salsa, Cuban son, and Dominican (and Haitian) meringue's emphasis on hard-driving syncopations and percussion breaks was adopted by early hip-hop pioneers.[28] The reason was in part because DJs played music primarily for dancers. B-boying or b-girling (described by the media as breakdancing) practically developed simultaneously with hip-hop (Figure 6). Although

FIGURE 6
B-girl Laneski with the group Majestic Rockers in New York City, 1985. Courtesy of the National Museum of American History.

the music Kool Herc played was bass-heavy like Jamaican music, he found that Latin-tinged music was better at getting the audience moving—adding another key component to the hip-hop sound.[29] As Richard Sisco says, "What propelled it was all the b-boy action. That was what drove the music forward."[30] From dancer Popmaster Fabel's perspective, "We all knew when and where the breaks of the songs were. Anticipating these musical crescendos turned us into dancing timebombs. Instantly bodies would all simultaneously drop. From top-rocks to floor-rocks the breaks assaulted our senses."[31] Since the instrumental breaks were so popular, DJs sought techniques to extend them.

DJ Afrika Bambaataa and the Zulu Nation

DJ Afrika Bambaataa and the Zulu Nation organization he cofounded with five others used hip-hop as a means to end violence in the Bronx and in so doing contributed to hip-hop's promotion and success. Afrika Bambaataa was himself a former gang member who was partially motivated by gang activities to create an alternative for youth. DJ Disco Wiz recalls, "To be a DJ in those days . . . it was very risky. People were going to jams and pulling out guns and shooting up the place; people were getting bum-rushed or being robbed."[32] DJ Charlie Chase adds that "a lot of bad things did happen. Not at every party—it's not fair to say that something always happened."[33] In hip-hop's early days Afrika Bambaataa deejayed at the Bronx River Houses (Figure 7). This location also became the headquarters for the Zulu Nation that Bambaataa cofounded.

FIGURE 7
Afrika Bambaataa outside the Bronx River Houses with cassette tape recorder, 1986.

The Zulu Nation utilized the structure and some of the functions of gangs to help absorb and pacify them. The Zulu Nation helped enforce and keep the peace at Bambaataa's parties. As MC Sha-Rock recalls, "If you went to a Afrika Bambaataa party, you expect to be safe, because nobody was starting any trouble at a Afrika Bambaataa/Zulu Nation party. That was a good atmosphere, and I think that's what a lot of groups wanted—people to feel safe and come in and just listen to the music and everything."[34] Most DJs had security crews that helped DJs promote themselves, have a better presence at parties, move equipment, and also prevent theft, violence, and spying on the records they used. As Kool DJ AJ noted, "You had to bring your gang for protection, because my system might have cost like $15000, which I paid out of my pocket, and if you didn't have no gang behind you, you might've lost your system in the South Bronx."[35]

Bambaataa saw the music and movement he termed hip-hop as an opportunity for Bronx youth and young adults to create a safer place for themselves.[36] Grandmaster Flash affirms that "Bambaataa played a major part. He took all the different cliques and transformed fighting against each other into a more positive energy. . . . The block party thing caused peace in the neighborhood."[37] Afrika Bambaataa and the Zulu Nation created safe havens for hip-hop parties in the Bronx, which made traveling there for parties less intimidating. Hip-hop spread to the other NYC boroughs and beyond as people took what they heard and saw in the Bronx back to their communities.

> Bam would be playing the break-beats and then would jump off and start playing some calypso, or playing some reggae, or playing some rock. I was like "What is Bam doin?" But Bambaataa's mind-set was that hip-hop was an open field of music.
>
> VAN SILK

Bambaataa and The Zulu Nation also popularized the term hip-hop as the name of the new type of music and related culture created in the Bronx. He says that "the phrase hip-hop came from Lovebug Starski, who used to use it in rhymes. Back then the music had no name, maybe Boi-oi-oing, or Be Bop. Pulling it together as a culture, the name hip-hop started with myself and the Universal Zulu Nation."[38] Hip-hop originally encompasses deejaying, emceeing, dancing (b-boying/b-girling, popping, locking, and rocking), and aerosol art. Giving them an overarching name gave the culture power. As Bambaataa said, "Hip-hop saved a lot of lives, and brought the unification of many different people together under the banner of hip-hop culture."[39] Afrika Bambaataa, known as the godfather of hip-hop, played a key role in hip-hop's success by helping to define the components that made it a culture and developing a civic initiative role for it.

Finding Unusual Grooves

Both DJ Kool Herc and Afrika Bambaataa were known and loved for the obscure records they played. It is notable that many of the records that they as well as other DJs used were records from their parent's generation. As previously mentioned, Kool Herc chose music he thought would keep the dancers moving, but as Kool DJ AJ explains, Kool Herc "used to play a lot of records that you couldn't buy in the store. He played a lot of James Brown, Melvin Sparks, The Incredible Bongo Band, Baby Huey, a lot of real interesting breakbeats."[40] One of DJ Afrika Bambaataa's contributions to hip-hop is the innovative way he believed that any music, any sound, was fair game for mixing. As a consequence, Bambaataa expanded people's idea of what hip-hop could sound like. As he describes, "My audience was the most progressive of all, because they knew I was playing all types of weird records for them. I even played commercials that I taped off the television shows, from Andy Griffith to the Pink Panther, and people looked at me like I was crazy."[41] To find his eclectic mix of records, he spent a lot of time in record shops. "I had a broad taste in sound, and I was checkin' all into the rock section or the soul sections or the different African sections or the sections of the Latin records."[42] This diversity opened the door of possibility for what DJs conceived of as resources for sampling, and Afrika Bambaataa is recognized for his contribution with the title "Master of Records." He motivated DJs to be creative in their choices and dig through record store bins for unusual and interesting breaks. Listening to diverse albums gave DJs an incredible knowledge of music history. The different and the obscure break that was the hit of the party was a highly coveted secret for DJs. Afrika Bambaataa recounts this precaution: "You could be playing records with your group, and other people is sneaking up, trying to get next to you, to see what you was doing. So I used to peel the labels off or put water on there and take the cover off so they couldn't see."[43]

INVENTION

The originators of hip-hop grew up in an environment of urban poverty that was rich with the influences of New York City DJs, Jamaican immigrants, and dancers. Kool Herc and Afrika Bambaataa were responsible for some of hip-hop's first distinctive characteristics. For approximately the next two years they were the primary competition for all DJs. The winner of competitions or battles was usually the DJ with the loudest sound system. In about two years, however, with the contribution of Grandmaster Flash and GrandWizzard Theodore's inventions, hip-hop would truly come together.

Quick Mixing and the Peek-a-Boo System

In the early 1970s, turntables were built for one thing: playing records. They were not designed to be stopped with fingers or spun backward, so aspiring DJ Grandmaster Flash had to tinker with them to meet his needs. He ended up innovating many of the techniques that hip-hop DJs use today. Grandmaster Flash, whose given name is Joseph Saddler, had a lifelong interest in electronics. As he explains,

In my early teens, I had this habit of wanting to know how the internal workings of things operated. . . . My mother decided to send me to technical school. And since I was so intrigued with the internal workings of these electronic items, I then started learning the actual technical terms for these particular items. . . . Early on in high school, we had assignments where we had to build amplifiers. So I was able to understand how you diagnosed circuits, how you use a signal generator or ohmmeter, you know?[44]

Like other aspiring DJs in the Bronx, Grandmaster Flash sought to build a bigger, louder sound system than Kool Herc's. He was able to use his knowledge of electronics to find the parts he needed and assemble his own sound system, using the natural urban resources that the Bronx provided. He notes, "[I] went to junkyards, abandoned car lots. I asked supermarkets for the big jugs they put pig guts in, to make cabinets for my bass speakers."[45] He adapted objects to suit his needs such as using traffic light sensors as tweeters and as he explains, "I had to take microphone mixers and turn them into turntable mixers. I was taking speakers out of abandoned cars and using people's thrown-away stereos" (Figure 8).[46]

FIGURE 8
Technics SL-1200MK2 turntable, about 1979, used by Grandmaster Flash. The Technics 1200 was the most popular turntable used by DJs. © 2014 Smithsonian Institution; photo by Hugh Talman. Courtesy of National Museum of American History.

One of Grandmaster Flash's most important contributions to the development of hip-hop dee-jaying was the idea that all surfaces of a record could be handled. Most people handled vinyl records by their edges, but Grandmaster Flash was unafraid to place his hands on the grooves and move the record how and where he wanted it.[47] He recalls, "to some degree I was ridiculed by all the other jocks that was doing the other style of DJ-ing, saying that I destroy records" (Figure 9).[48]

Physically manipulating the records was a real breakthrough, but Grandmaster Flash was ultimately seeking an efficient technique for creating an endlessly looped instrumental break. He elaborates, "To every great record, there's a great part. This is what we used to call 'the get down part.' This is before it was tagged 'the break.'... And this particular part of the record . . . unjustifiably, was maybe five seconds or less. . . . So in my mind, in the early seventies, I was picturing, 'Wow, it would really be nice if that passage of music could be extended to like five minutes.'"[49] He worked to find a solution for years, saying "I sacrificed most of my kid years to try to just take this thing that was just running around in my head and make it a reality."[50]

With the two turntables and a mixer setup, DJs had two copies of the same record, one on each turntable. They could have one record cued up to the start of the instrumental break on one turntable while it was playing on the other turntable. Using the slip-cueing technique, when the instrumental break ended, they would use the mixer to start the cued-up break on the other turntable,

FIGURE 9
Grandmaster Flash at the turntables, 1970s.
Photograph © Ernie Paniccioli.

FIGURE 10
"Bustin' Loose" record by Chuck Brown and the Soul Searchers, 1978. Grandmaster Flash used this record while deejaying. Chuck Brown's "Bustin' Loose" was a common record for DJs to sample from. The grease marks he made to mark his record for quick mixing are visible on the record's grooves and label. © 2014 Smithsonian Institution; photo by Richard Strauss. Courtesy of National Museum of American History.

thus looping the instrumental break. To easily loop the breaks, Grandmaster Flash invented the now fundamental deejaying technique of quick mixing. As historian Mark Katz explains, "Taking a grease pencil he drew one line from the center hole to the edge of the label to indicate the beginning of the break, and another to show where it ended. . . . With the label facing up, he treated the cardinal points of the record as twelve, three, six, and nine o'clock. His pencil marks were like hour hands, and he could see at a glance that a break, say, started at two and ended at ten" (Figure 10).[51]

Quick mixing is now a quintessential part of deejaying, and so is the iconic peek-a-boo system Grandmaster Flash developed. Although there were mixers available with built in cue systems, they were typically expensive. DJ Grandmaster Flash made his own system that he termed the peek-a-boo. As he explains in several interviews,

> I couldn't afford a mixer with a built-in cue system where you could hear turntable one or two in advance.[52]
>
> I had to go to the raw shop parts downtown to find me a single pole double throw switch, some crazy glue to glue this part to my mixer, an external amplifier, and a headphone.[53]
>
> I'd cut the wires from turntable jacks and take the two and turn it into four. I'd take the two and run it into an 8-watt external amp—just enough to power a headphone—but I'd run it into a single-pole, double throw switch and I would wire it ground-to-positive, so when you clicked it this way I could hear what this turntable was doing before I pushed the fader up- or vice versa. It allowed me to "peek" into the signal.[54] (Figure 11)

Grandmaster Flash's contributions of the peek-a-boo system, quickmixing technique, and fearlessness to manipulate the records were added to the sound developed by Kool Herc and Afrika Bambaataa, redefining and progressing the art form of the hip-hop DJ.

Scratching

One of hip-hop's signature sounds is scratching—the rhythmic scratch of a record. Bronx native GrandWizzard Theodore is known as its inventor. Theodore Livingston grew up in hip-hop. Twelve-year-old Theodore would help carry equipment, set up speakers, and pull out records to play at parties held by his older brother DJ Mean Gene, of the L Brothers, who was Grandmaster Flash's DJ partner. During this time GrandWizzard Theodore became an expert at dropping the record needle in just the right place for each song's break but was "trying to think of new ways

FIGURE 12
DJ GrandWizzard Theodore
at the T-Connection, 1980.
Photograph © Charlie Ahearn.

to be different. I wanted someone to close their eyes and say, 'That could only be one person—GrandWizzard Theodore—playing that record'" (Figure 12).[55]

Legend has it that Theodore was at home making a mix tape to play at school when his mom opened the door. As she told him to quiet down, Theodore kept one hand on the record, causing the needle to make a scratching sound while the other record was still playing. Theodore liked the sound:

> What I did was give [the rub of a record] a rhythm; I made a tune out of it, rubbing it for three, four minutes, making it a scratch. Before that a DJ would have

his earphones on, and I could see him moving the record back and forth, but the people couldn't hear him moving it back and forth. What I did was, I just switched it over so the people could hear the record going back and forth.[56]

Then he "practiced with it and perfected it and used it with different records, and that's when it became a scratch."[57]

Conceiving the scratch of a record as a musical element was a huge moment in hip-hop history—it became the signature sound of hip-hop deejaying. It also offered DJs another creative outlet within deejaying. Grandmaster Flash and GrandWizzard Theodore were major contributors to what would become known as turntablism (using the turntable as a musical instrument).

Technological Elements of Hip-Hop Unite

With more inventions and innovations to come, the foundational elements of hip-hop culture were established. The influence of different aspects of the Bronx can be seen in each invention. Kool Herc was most influenced by the Bronx's ethnic character, Afrika Bambaataa by its problems, Grandmaster Flash by its urban resources, and GrandWizzard Theodore by hip-hop itself. Their contributions combined to create more than just a music genre—they defined what a DJ could do.

As hip-hop deejaying developed, it influenced the design of sound and recording equipment, the turntable, records, headphones, and mixer. In the early days of hip-hop, equipment was often not designed to create the effects that DJs wanted. Overused stiff faders would break or create static, needles were more prone to skip, and most mixers did not have built-in cue systems. "The equipment wasn't made for what we wanted to do with it," Mercedes Ladies DJ Baby D remembered, "so we urbanized it."[58] DJs adapted equipment through methods such as weighing down tonearms with pennies, building their own cueing systems, finding alternative ways to mix records without mixers, using grease to loosen faders, and creating their own slip mats out of whatever material they could find. Even into the late 1980s mixers were not designed for using scratching and quick mixing techniques. Modifications for hip-hop DJs were eventually incorporated into the designs of new models of mixers and turntables and have become standards for contemporary deejaying equipment (Figure 13). Innovations in music sampling technology that made it easy to record and store tracks made it even easier for DJs to create music (Figure 14).

Invention in Every Space

DJs from the Bronx used every corner of their borough as venues to showcase and test their innovations: schoolyards, school gyms, abandoned buildings, streets, parks, after-hours clubs, houses, community centers, rec rooms in housing projects, police athletic centers, and even a Burger King (Figure 15).[59] Like the Apple and Medtronic founders who adapted garages to be workshops, Bronxites transformed suitable spaces into temporary performance venues. DJs and their crews

FIGURE 13
Grandmaster Flash model Rane
Empath DJ mixer, about 2003–06.
Grandmaster Flash worked with
Rane Empath to produce this mixer
designed for the needs of hip-hop
DJs. © 2014 Smithsonian Institution;
photo by Hugh Talman. Courtesy
of National Museum of American
History.

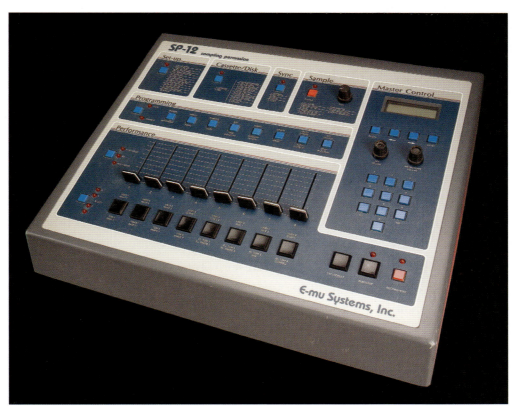

FIGURE 14
E-mu Emulator (drum sampler), 1985–86. Samplers became a primary tool for many hip-hop
DJs and producers because they could be used instead of records and turntables for sampling.
Samplers could record and store sound, manipulate it, and play it back. © 2014 Smithsonian
Institution; photo by Hugh Talman. Courtesy of National Museum of American History.

G Man and his crew deejaying at a park, 1985. Photograph © Henry Chalfant.

hauled their equipment and records from one venue to the next. Outdoor venues presented lighting and power problems for holding parties during the evening, but people found solutions like connecting light bulbs to their equipment for light. Power could often be borrowed. According to Busy Bee, "[Mean] Gene would show up and ask someone, 'We'll be here playing a little music. We can give you $30, if we can plug into your house for three hours.' Run the extension cord out their window, we on!"[60] Many DJs would hot-wire streetlamps. As Grandmaster Flash describes, "I'd break open a faceplate on the light pole, and thanks to what I'd learned at vocational school, I could split the wires, step the power down, and make it all work."[61] It was a sight to behold, as Nelson George recalls:

> The sun hadn't gone down yet, and kids were just hanging out, waiting for something to happen. Van pulls up, a bunch of guys come out with a table, crates of records. They unscrew the base of the light pole, take their equipment, attach it to that, get the electricity—Boom! We got a concert right here in the schoolyard and it's this guy Kool Herc.[62]

The lack of police enforcement was a distinct advantage for these outdoor jams. MC Busy Bee of the L Brothers characterized his encounters with police who enabled outdoor parties to carry on:

> The police didn't really know about hip-hop at that time. They knew what time to tell us to stop, they knew that much, but as far as bothering us or harassing us, they didn't do it. Come about 10 o'clock, or close to 11, they would come around. "You want us to turn it off now?" "Yeah." Or, "No, turn it down just a little bit. We'll give you another hour or two." You know? That's how we did it. We cooperated with them, it wasn't like, "Cut that off!"[63]

Indoor venues required workarounds too; as MC Busy Bee explains,

> Early on, we did the streets, the backyards, the schools, and stuff like that. The janitors—that's how we got into the schools, because the janitor cleans up when school is out. So if you can find a janitor that was cool, you can say, "Hey listen, I've got twenty extra dollars for you to let us use the gym at night, and we'll clean it up after night."[64]

More popular DJs eventually had steady gigs at clubs (Figure 16). All of these venues were key spaces for the evolution of hip-hop.

Hip-hop audiences were just as important as DJs for inventing hip-hop. Competition to create a system that rivaled Kool Herc's created new innovations in sound systems, but competition for getting the crowd going also drove DJs to find the best breaks and encouraged innovations in new techniques to wow the crowd. Like a "mad technoscientist," Grandmaster Flash never stopped working and trying to progress: "I had spent months holed up in my room testing dozens of needles, sampling sounds and perfecting my newest experiment."[65]

> Sometimes it's hot and a lot of the clubs didn't have no air conditioner. So I gave parties out in the park to cool out while summer's there. To play in the park is to give the fans and the people something.
>
> DJ KOOL HERC

DJs interacted with the audience and their competitors, motivating them to innovate. DJs more or less created what they thought the audience would like and selected their sounds, samples, and breaks accordingly. As Grandmaster Flash explains, "You can actually test your newfound jams to see how they work on the public right then and there. You'll know right then and there if you got something, as soon as you play it."[66] The audience was an important sounding board for hip-hop

FIGURE 16
The crowd below the DJ booth at the T-Connection (a popular club), 1980. In the DJ booth are GrandWizzard Theodore, Ritchie T, and Kool Herc. Photograph © Charlie Ahearn.

DJs. It often had the most say during DJ battles, which in hip-hop's early days were more like sound clashes.

For a structured contest a DJ would challenge another to meet at a specific time and place for a battle.[67] Most had a crew to give them an aura of power and intimidation (and because DJs had so much large and heavy equipment to transport to and from gigs, crews helped transport it and ensure that it was not stolen).[68] DJs set up their equipment on opposing sides of the venue, and the one with the biggest sound and best crowd response typically won.[69]

In early hip-hop, most battles were sound clashes that were won by drowning out the competition with the loudest system. As Disco Wiz says, "A lot of times guys won and they were wack DJs, but they had an awesome sound system and would drown you out."[70] However, after Grand-master Flash came on the scene and DJs perfected their individual styles, skill and technique increasingly decided the outcome.[71] "Kool Herc couldn't draw a crowd after people saw Flash,"[72] said

DJ AJ. He theorized that "it was like Flash had skills and Herc had the records, but maybe people was really more into the skills because they had heard all the records already."[73]

HIP-HOP BEYOND THE BRONX

Hip-hop spread from the Bronx to other boroughs of NYC. In addition to word of mouth, DJs promoted their parties with flyers that were sometimes designed by aerosol artists (Figure 17). Also, recordings made during parties and battles were sold, spreading the music around the city. "Around 1977," recalls DJ Disco Wiz, "we used to record all our battles. Every party we had we always had a boom box on the side, and we used to record what we did and who we did it to (Figure 18). And those things used to sell—we used to sell them in high school."[74] "We were selling cassettes of our mixes that were really our first albums," remembered Afrika Bambaataa.[75] Soon Bronx DJs started playing in other parts of NYC. As hip-hop traveled, it continued to evolve. At shows, the MC began to take the place of the DJ as the front man.

The song "Rapper's Delight," recorded and released in 1979 by the Sugar Hill Gang, introduced hip-hop to the rest of the United States. Soon it was taken up by a new generation. According to

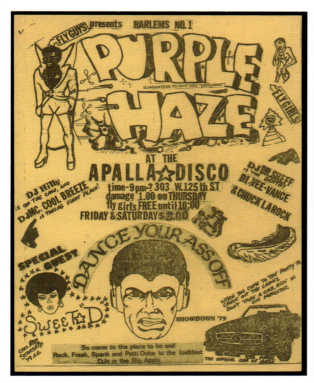

FIGURE 17
(Left) Purple Haze hip-hop flyer, 1979. (Right) Holloween Disco hip-hop flyer (date unknown). Early hip-hop flyers were an instrumental source of information about where to see hip-hop and also to promote DJs. Courtesy of the National Museum of American History.

FIGURE 18.
Sharp Multi-purpose Music System HK-9000
boom box, about 1985. Boom boxes brought hip-hop
to anyone in hearing distance. This one belonged to Fab 5
Freddy, hip-hop visual artist, filmmaker, and former host of
Yo! MTV Raps. © 2014 Smithsonian Institution; photo by Hugh
Talman. Courtesy of National Museum of American History.

the Cold Crush Brother's GrandMaster Caz and Charlie Chase, "Run-DMC was the cutoff point between us and hip-hop after that. . . . Because we had taken hip-hop from its bare necessities, from a baby, from being just in the park with speakers and stuff and plugging into light poles, up to doing shows in leather outfits."[76] "They took hip-hop to that other level. You could actually say they picked up where we left off. There was a next level there, and I guess we just couldn't do it."[77]

In the 1980s, DJ techniques expanded geographically, technologically, and in diversity. Grand-mixer D.ST's improvisational scratching on Herbie Hancock's song "Rockit" introduced the turntables as a musical instrument to the mainstream.[78] Artists from all elements of hip-hop culture began touring (Figure 19). Hip-hop hot spots emerged all over in places like Philadelphia and Los Angeles and across the world in London, Paris, and beyond. As samplers became increasingly central to hip-hop, the role of the DJs and producers began to overlap, and DJs expanded into producers' territory. In the late 1980s and early 1990s, hip-hop became more mainstream in part by having a more pop-friendly sound and mediums such as MTV. Today, hip-hop culture is global in scale and diversity.

The Bronx is economically revitalizing and remains a place where hip-hop culture flourishes, but it is also a place whose hip-hop history is celebrated. The Bronx's regeneration began in the mid-1980s when New York's City's government and two federal government programs gave money and expertise to existing Bronx nonprofit community development corporations.[79] They

FIGURE 19
Grandmixer D.ST in London, 1982. This performance was during the first hip-hop tour to stop in London. On tour were Afrika Bambaataa, D.ST, Rock Steady Crew, Fab 5 Freddy, Double Dutch Girls, and many more. Photograph © Janette Beckman.

collaborate to provide residents of a variety of income levels with housing and community services such as libraries and child care.[80] They also encourage home ownership, investment by banks, and direct management of city-owned buildings.[81] Crime has decreased, new buildings stand on the plots of derelict ones, and some big chain stores have set up shop.

Although the explosion of hip-hop innovation in the Bronx has tapered, you can still attend hip-hop performances in the Bronx with both contemporary artists and older ones. Tools of War grassroots Hip Hop recreates the park jams of the 1970s and early 1980s at Crotona Park in the Bronx, exclusively featuring hip-hop DJ pioneers and DJ legends, including Afrika Bambaataa, GrandMaster Caz, GrandWizzard Theodore, DJ Jazzy Jay, and Kool DJ Red Alert. Hip-hop history tours of the Bronx are offered and are popular, especially with international visitors. Several hip-hop artists are featured on the Bronx Walk of Fame. In celebrating its hip-hop history the Bronx remains a hot spot for inspiring future DJs who may continue to innovate. In the 1970s the Bronx and the people who lived in it had the right combination of resources and background to innovate a new music, a new musical instrument, and the techniques for playing it that would change music forever.

PLACES OF INVENTiON

Affiliates Project

THE RISE OF SEATTLE'S GAMING INDUSTRY
(1980s–TODAY)

Anna Karvellas

Greater Seattle *Places of Invention* Team:
The Museum of History & Industry
with 20after1

In the past ten years, Seattle's gaming industry has grown from 37 companies to over 330. In 2013 alone, these companies brought in revenues over $19 billion.[1] Why is Seattle the center of game innovation? Working with the Greater Seattle game community, a history museum documents how place, people, and circumstance came together to spark the region's gaming culture.

THE MUSEUM AND SOUTH LAKE UNION

A place of invention within a neighborhood of invention within a region of invention, the Museum of History & Industry (MOHAI) lives its mission to use invention and innovation as a lens for understanding Greater Seattle history (Figure 1). Thoughtfully constructed exhibitions explore ways that the physical landscape has shaped—and been shaped by—its people. They follow Seattle's transition from frontier town to port city to center of the aviation world; its prominence in the tech and creative industries; and the thriving fields where these two industries meet, most notably in gaming. Names like Boeing, Microsoft, and Amazon hover like so many cloud-computing systems against the backdrop of Puget Sound, Lake Washington, Mount Rainier, and the 1962 World's Fair tribute to innovation—the Space Needle.

OPENING IMAGE (OPPOSITE PAGE)
South Lake Union with MOHAI in the distance. Photograph by Chris J. Gauthier. © 2014 Smithsonian Institution.

FIGURE 1
Museum of History & Industry, South Lake Union, Seattle, Washington, 2014.
Photograph by Kathleen Kennedy Knies. Property of MOHAI.

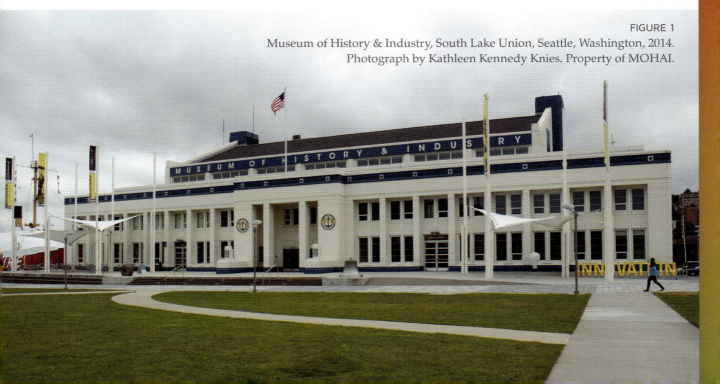

The Museum of History & Industry's participation in the *Places of Invention* (*POI*) Affiliates Project coincided with its move to a new space designed to allow greater focus on local innovation. The museum reopened in the former U.S. Naval Reserve Armory Building located in South Lake Union, a neighborhood rapidly transforming into a life science and technology hub (opening image). Collectively, businesses in the former warehouse district generate $3 billion in sales.[2] Spurring the development is Amazon's construction of an expanded world headquarters and the necessary infrastructure to support its workforce.

The Museum of History & Industry's location allows its staff to connect historic and contemporary innovation in new and interesting ways. Dr. Lorraine McConaghy, MOHAI public historian and *POI* project consultant, has been deeply involved in curating the museum's new exhibition spaces, including the Bezos Center for Innovation.[3] Many of the same principles and strategies found there can be found in the *POI* Affiliates Project that she so directly inspired. More on this work can be found in this book's concluding essay, "What's Next?: Museums and Innovation" (Figures 2, 3, and 4).

FIGURE 2
Dr. Lorraine McConaghy leads a tour through MOHAI's Bezos Center for Innovation, Seattle, Washington, 2013. Photograph by Daniel Sheehan. Property of MOHAI.

FIGURE 3
Visitors exploring Seattle-based innovation and its connection with the world, Bezos Center for Innovation, MOHAI, Seattle, Washington, 2013. Photograph by Daniel Sheehan. Property of MOHAI.

FIGURE 4
Interactive "patent wheel" of Seattle inventions, Bezos Center for Innovation, MOHAI, Seattle, Washington, 2013. Photograph by Daniel Sheehan. Property of MOHAI.

SELECTING A PLACE-BASED INVENTION TOPIC

When MOHAI's Helen Divjak and 20after1's Sean Vesce arrived at the Smithsonian's training workshop, they had already decided to focus on gaming innovation in Greater Seattle. As a designer and developer of original and licensed video games, Vesce had seen the growth of Seattle's gaming industry first-hand. There are more game companies in the region than anywhere else in the world. According to a recent enterpriseSeattle report, over 16,500 employees work directly in Seattle's game industry, with an estimated 80,000 people working in occupations suitable or transferrable to interactive media.[4] The region is home to established and independent game companies, and its share of software developers is 3.2 times that of the national economy (Figures 5a and 5b).[5]

FIGURE 5a

Game Industry in Seattle Area, 2013, Innovation Partnership Zone brochure. Washington Interactive Network. Used with permission.

A lot of people like to think that Silicon Valley carries some of that weight, but the reality is that this is the epicenter of gaming, whether it's casual or core or mobile or PC or console. . . . The community here is alive in spades. . . . There is no doubt that this is ground zero for games and will continue to be for a long time.[6]

—John Holland, Big Fish Games

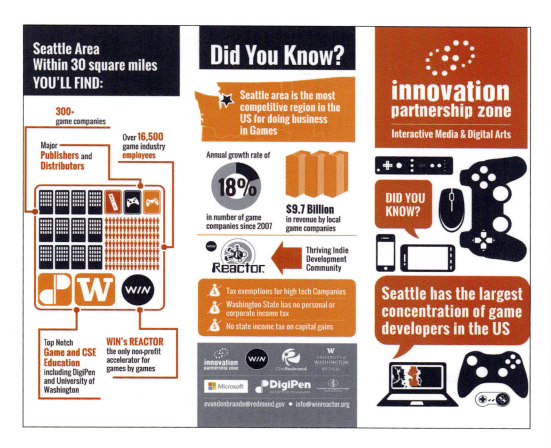

FIGURE 5b

Seattle is a center for cultural arts and technology, and games is really about how those come together in a unique way.[7]

—Kiki Wolfkill, 343 Industries

The Museum of History & Industry and 20after1 wanted to use *POI* as an opportunity to document an important, and lesser known, part of the Seattle innovation story, but it was more than that: invention and innovation in the gaming industry is happening *now*, and they had an opportunity to explore this dynamic ecosystem in action. The team hoped to capture some of that zeitgeist and show visitors how innovation builds upon itself by using a story that touched many lives across multiple generations and cultures. The Museum of History & Industry was especially interested in looking at gaming as a distinctive basket of innovation in which designers invent and create across the board—from story, art, and music to software and hardware.

FOCUSING THE LENS: WHY THERE? WHY THEN?

Vesce's experience in the creative and technical sides of the gaming community allowed him to connect Divjak's successor at MOHAI, Julia Swan, with some of the field's most influential people.

Swan conducted in-depth oral history interviews with seven individuals who have made considerable contributions to Greater Seattle's gaming industry:

- Ed Fries, former vice president of Microsoft Game Studios and founder of the Xbox project
- Megan Gaiser, founder of Contagious Creativity and former chief executive officer and chief creative strategy officer of Her Interactive
- Richard Garfield, game designer and creator of the game Magic: The Gathering
- Jerry Holkins, writer, and Mike Krahulik, cartoonist, *Penny Arcade* Web comic and founders of the Penny Arcade Expo (PAX), the world's largest annual gaming convention
- Kim Swift, former creative director of Airtight Games and game designer for Valve Corporation where she co-created the game Portal
- Paul Thelen, founder, chairman, and chief executive officer of Big Fish Games

As part of the *POI* program, edited and unedited versions of each interview are available in perpetuity through the MOHAI and Smithsonian archives. Transcriptions are also available for download directly through the Lemelson Center website.[8]

Over eleven hours of video interviews, the seven participants were asked questions pertaining to *POI* core-concepts. They spoke about the region's software and tech origins, the influence of programs at DigiPen and the University of Washington, the impact of business tax credits, the lack of income tax, and other key factors. Most significantly, they talked about the exchange of ideas within a highly entrepreneurial and collaborative community willing to take strategic risks in pursuit of something new or improved. For the *POI* team, these conversations reinforced the importance of certain behaviors common to any place of invention while elucidating the ways in which Seattle's particular combination of place, people, and circumstances allows gaming innovation to thrive. Excerpts from the interviews form the heart of MOHAI's mini-documentary on Seattle gaming for the Smithsonian exhibition. The interviews also led to a *POI*-themed panel at PAX Prime and a MOHAI exhibition on the past, present, and future of gaming guest-curated by Swift. (See the "*Places of Invention* Projects and Programs" section for more on these programs.) Through the project, the practices of gaming invention and innovation were examined and modeled at MOHAI with active practitioners (Figure 6).

The Common Denominator: Microsoft's Tech Ecosystem

> In 1962, Seattle hosted a forward-looking World's Fair that helped brand the region as a center for big ideas. So it's no coincidence that Seattleites like Bill Gates and Paul Allen decided to bring their company, Microsoft, back to their home state in 1979—a move that inspired the growth of an entire industry of high technology and set the stage for Seattle to be a place of invention in gaming.[9]
>
> —MOHAI *POI* video

In the course of the interviews, certain drivers and relationships between geography, community, and resources emerged in Seattle's gaming innovation story. One common denominator appeared in every oral history. Microsoft and the "tech ecosystem" it created when it moved to the region. (Figures 7–10):

I think why Seattle is becoming a big [game] technology hub is honestly Microsoft. The game industry kind of started to spring up around Microsoft: people making games for Windows, people making games for the consoles—like for the 360—a lot of game companies in the area are started by ex-Microsoft employees. Gabe Newell [of Valve] was an ex-Microsoft employee. Ed Fries, who is a partner in my company, is ex-Microsoft. My other two partners, Matt Brunner and Jim Deal, are also ex-Microsoft. So it's one of those things where Microsoft is this hub that things have started to revolve around, starting about twenty years ago.[10]

—Kim Swift

FIGURE 7
Kim Swift, game designer, MOHAI, Seattle, Washington, 2013. National Museum of American History. © 2013 Smithsonian Institution.

What happens in the game business is that people start companies and they grow and they work on things and then people split off and then they start new companies and they grow and they work on things and then people split off and then they start new companies and it just keeps fracturing and fracturing. And now it's kind of complicated to trace the lineage all the way back. The one core thing I think you would find [is] . . . the Microsoft connection for most people.[11]

—Ed Fries

FIGURE 8
Ed Fries, former vice president of Microsoft Game Studios, MOHAI, Seattle, Washington, 2013. National Museum of American History. © 2013 Smithsonian Institution.

If you look at Seattle's history and you focus all the way back to just tech in general, Microsoft is a massive company and it attracts a ton of really talented people to the area. . . . There are a lot of people passionate about games [and] there is a generation of gamers that now have technology degrees. Microsoft [is] one of the big concentrations of those technical people—people saying, "I want to start a company and I want to be passionate about it," and games is a natural fit.[12]

—Paul Thelen

FIGURE 9
Paul Thelen, founder of Big Fish Games, MOHAI, Seattle, Washington, 2013. National Museum of American History. © 2013 Smithsonian Institution.

The biggest reason has to be Microsoft—the tech community around Microsoft has gone off and started all sorts of companies, and some of them are computer games, and some of them are like Cranium. . . . Gaming is intrinsic in some ways in the tech industry.[13]

—Richard Garfield

FIGURE 10
Richard Garfield, game designer, MOHAI, Seattle, Washington, 2013. National Museum of American History. © 2013 Smithsonian Institution.

Liquidity

Also advantageous is the liquidity of Seattle's gaming community—the way, as Paul Thelen points out, individuals can move more easily from job to job in an ecosystem large enough to allow for different career paths in the same industry.[14] Besides job mobility, there is also the attraction of solving new challenges with other creative people drawn to the region. Similar factors have been critical to the successes of other places of invention examined in this book, including Silicon Valley, Fort Collins, and Medical Alley.

> Some of it is just getting enough of a community that the community kind of becomes self-supporting and self-generating. So I think Seattle offers kind of a nice environment, physically. It's not ridiculously expensive. It's got a good tech infrastructure. It's a place that is recognized as one of the nicest cities to live in in the country. It's just got a bunch of different intangibles that all come together to make it a nice place. And then one of the biggest things is that you've got this history. That game development has been happening here a long time. And teams split, and you've got new teams, and they've got families and friends and they don't want to move out. And so it just kind of grows organically.[15]
>
> —Ed Fries

> Lots of people will leave [games like] Wizards and go work for Microsoft. They'll work in their games division and bring different game ideas and then they'll be frustrated there and they'll leave and start a start-up firm. So you'll see these people—the paper game experts which come out of Wizards, and the variety of companies like Whiz Kid [Games] that have spawned after Wizards, and the broad technical excellence of Microsoft, and now Google, and Amazon, and all these others—I'm sure that mixing pot is a really important reason. It feeds on itself.[16]
>
> —Richard Garfield

Jerry Holkins and Mike Krahulik, creators of *Penny Arcade* (Figure 11), agree:

> Holkins: [The gaming industry] is a huge part of why Penny Arcade has stayed here.
>
> Krahulik: It's huge. The reality of it is that we can do it from anywhere. We tried to tell [our business manager] Robert we should do it from Hawaii. But they don't have a gaming community.

FIGURE 11

Mike Krahulik (left) and Jerry Holkins (right), creators of *Penny Arcade*, MOHAI, Seattle, Washington, 2013. National Museum of American History. © 2013 Smithsonian Institution.

Holkins: And that's it. It has specifically to do with the interplay between all the people that like games, and a lot of people making games. Innovative games. And even a game platform is a local creation. I think it has to do with the particular ecosystem here. Because you could run a show like PAX and it'll sell out in three hours. Your show that takes place in the state's largest convention hall, will sell out in three hours. You don't get that in every state.

Krahulik: Absolutely that's why we're staying. . . . I mean, obviously, having Microsoft here is incredible. Having Nintendo here is incredible. You have all these super smart creative people coming in to work at those places, and the . . . branching out—like they have their own idea for a game, so they are going to stay in Seattle.

Holkins: And they are here already, and the network that they have built. What in another era would have been called their Rolodex is pretty deep here. There are a lot of people to know here that can enable creation. There are a lot of structures of support.[17]

Community

Like any place of invention, community plays an important role in Seattle's gaming culture. Mentorship programs and networking opportunities are a part of Seattle's game culture, most notably through the Washington Interactive Network (WIN), started by Kristina Erickson Hudson.

[I met] Kristina Erickson Hudson [when] she was working for the film board, not long after I left Microsoft in 2004. She was trying to understand the video game business along with some friends. I showed them around the Game Developers Conference one year and I talked about the fact that I know all these local game companies because I've worked with them in my job but a lot of times they don't know each other. I was holding some dinners just to bring people together. Sometimes their offices would be across the street from each other but they didn't know each other. So I'd have these dinners with different head of companies so that they could meet.

WIN was kind of a formalization of that. Kristina ended up leaving the film board and went to enterpriseSeattle. . . . Through them and now through a government grant she's been able to start this WIN project. It holds monthly meetings to get people together to talk about different topics: starting game companies, running game companies. [She's] been able to start an incubator here in Seattle to start new game companies—a group called REACTOR—and run an annual event called Power of Play, which we just did about a month ago. That's been one great way to get people together and get them to know about each other.[18]

—Ed Fries

At Her Interactive, game producer Megan Gaiser and her colleagues made a conscious decision to involve girls and women in the game development process. They also sought to actively nurture the interest of future game developers through office tours and visits with female designers, marketers, and executives.[19] For Gaiser (Figure 12), the ability to connect with others this way is part of Seattle's broader appeal.

I like the openness of Seattle. Maybe because there are so many game developers, collaboration is really natural, I've found. There are a lot of resources available to you as well. When I got the opportunity to lead the company I realized there were a lot of skills I didn't have. And I took it upon myself to call different CEOs here and in San Francisco and introduce myself and see how we could collaborate, and for the most part, people were incredibly open and I forged some really strong relationships. We would share information with each other as we were learning. That helped me a lot. I found mentors and became a mentor to several people. And I think that is very important. . . . In general, this is a very creative field,

FIGURE 12
Megan Gaiser, founder of Contagious Creativity, MOHAI, Seattle, Washington, 2013. National Museum of American History. © 2013 Smithsonian Institution.

and things change very quickly. Part of the requirement to succeed in gaming is to be nimble and flexible and adaptive and resourceful. And scrappy. And to do that, you need to be learning constantly. The broader your network is, the better chance you will have to succeed. It is not an industry where you can be in your little room making your games, because so much is happening now—technology is changing, markets are changing.[20]

—Megan Gaiser

And What About the Rain?

Seattle is more than just the home of PAX Prime. The greater Seattle region is a major hub of the gaming industry. But how did that happen? Is it really because of the rain?[21]

—PAX Prime *POI* panel description, 2013

Julia Swan opened the *POI* panel on Seattle gaming with this question, and the more than 300 people in the packed room laughed knowingly. Although Seattle and rain are an enduring stereotype (Figure 13), the truth is, Ed Fries remarked, that you just feel less guilty staying inside playing and designing games when it's raining outside. When other people you know also are inside playing and designing games, coming up with new ideas and products—maybe challenging you as a player, designer, programmer, or producer—there is even more enticement.

There's not a whole lot of reason to go outside here—most of the time—and I don't—I honestly don't think you can discount that as being a part of this structure.[22]

—Jerry Holkins

Staying in and playing on your computer is pretty good most of the time here. . . . You want to have fun. When you want to play things that are fun inside when it is raining outside, you do that here.[23]

—Mike Krahulik

So you get something big like Microsoft in some area like Seattle, where there is a lot of reason to be inside some of the year—and a lot of reason to be outside during the beautiful summers. It's a good formula for games.[24]

—Richard Garfield

PLACES OF INVENTION PROJECTS AND PROGRAMS

The Museum of History & Industry's investigation into Seattle as a place of gaming innovation aligns well with its ongoing work with the Seattle tech industry. In addition to the oral histories, a growing number of *POI*-related programs have taken place. One *POI* project informs another at MOHAI and vice versa. The result is an expanding body of documented work available through MOHAI and Smithsonian archives.

Places of Invention Interactive Map Videos
Gaming in Greater Seattle

After a brief history of Seattle innovation, this video features excerpts from MOHAI's interviews with Ed Fries, Megan Gaiser, Richard Garfield, Jerry Holkins, Mike Krahulik, Kim Swift, and Paul

Thelen. Themes emerge as the subjects talk about their experiences in the field and the people, resources, and circumstances that made Seattle a center of gaming invention and innovation. The video is available through the *POI* interactive map and through various MOHAI channels.

Past, Present, and Future of Video Games
Exhibition at MOHAI's Bezos Center for Innovation, Guest Curated by Kim Swift

Kim Swift's relationship with MOHAI continued beyond her oral history interview in April 2013. She developed one of the inaugural exhibits for its Bezos Center for Innovation. In the exhibition and a related curator's talk, Swift explored the history of video games and the industry's future. After viewing games from different eras, visitors were encouraged to think about the ways that video games are inventive and innovative and whether their pros outweighed their cons. Opinions could be posted and shared on a community wall in the exhibition (Figure 14).

A Byte of Seattle: The Rise of Seattle's Gaming Industry
POI-Themed Program at PAX Prime, the World's Largest Gaming Convention, Washington State Convention Center, Seattle, Washington, 2013

A panel on MOHAI's *POI* work with the Smithsonian to explore the history of gaming innovation in Greater Seattle was presented at PAX Prime.[25] The moderator was Julia Swan, and panelists were

FIGURE 14
Video game pro/con board with handwritten
visitor responses, Bezos Center for Innovation,
MOHAI, Seattle, Washington, 2014. Photograph
by Daniel Sheehan. Property of MOHAI.

Ed Fries, Kim Swift, Andrew Perti (the founder of the Seattle Interactive Media Museum), and Anna Karvellas (Figures 15 and 16).

> You can't forget that Seattle is home to the epic Penny Arcade Expo, better known as PAX, that attracts more than 70,000 gaming fans who come together to play anything from Super Mario on NES to Watch Dogs for Xbox One, hear from industry leaders on various panels and compete in tournaments.[26]
>
> —Taylor Soper, GeekWire, 2013

Seattle Anti-Freeze: Games
MOHAI, Seattle, Washington, 2013

The Seattle *POI* team created a *POI*-themed games evening at one of the museum's monthly social events (Figure 17) designed to bring people together around a topic.[27] Julia Swan presented a short history of Seattle gaming focusing on five critical factors in its development: Microsoft; Nintendo of America/DigiPen Institute of Technology; board games such as Cranium and Magic: The Gathering; the rise of mobile and casual games; and the Penny Arcade Expo (PAX). After the presentation, guests were invited to play video and board games that were designed by Seattle area game designers, for example, Halo on the original Xbox and Super Mario Brothers on a Nintendo Entertainment System. Board games were represented, too, with local game shop staff on hand to teach guests how to play Magic, Roll for It!, Guillotine, or King of Tokyo.

FIGURE 16
Scenes from PAX Prime, Seattle,
Washington, 2013. Gamers try out
new products on the trade floor.
Photographs by Anna Karvellas.
© 2013 Smithsonian Institution.

THE PAX 10

100+ submissions from independant developers across the globe. 50+ game industry judges. We proudly present to you the PAX 10 for 2013

Avalanche 2: Super Avalanche
Beast Games

BADLAND
Frogmind

Escape Goat 2
MagicalTimeBean

Gunpoint
Suspicious Developments

Lovers in a Dangerous Spacetime
Asteroid Base

Owlboy
D-Pad Studio

Ridiculous Fishing
Vlambeer

Rogue Legacy
Cellar Door Games

Sokobond
Alan Hazelden, Harry Lee, Ryan Roth

TowerFall
Matt Makes Games

Special Thanks and Consideration Goes To

DigiPen

FIGURE 17
Seattle Anti-Freeze presentation, MOHAI, Seattle, Washington, 2013. Property of MOHAI.

After-School Video Game Design Club (in partnership with E-Line Media)
MOHAI, Seattle, Washington, 2013

As part of an after-school club, students met once per week for six weeks to design, play, and share unique video games based on Seattle history.[28] Led by industry experts, the club engaged students using twenty-first century skills while inspiring their ingenuity and creativity. E-Line Media is a publisher of game-based learning products and services that "engages, educates, and empowers" students in science, technology, engineering, and mathematics (STEM) topics. *Places of Invention* community partner Sean Vesce is currently E-Line Media's creative director and studio manager. In 2015, MOHAI is partnering with E-Line Media to host a second videogame workshop for kids.

Kids who participated in After-School Video Game Design Club at MOHAI got to explore historic content in a new and exciting way, one that is relevant to them and their interests, while building skills in game mechanics and technology. Seattle has such a rich culture of gaming expertise to draw from, and E-Line Media was an exemplary partner.[29]

—Tara McCauley, MOHAI

MEDICAL ALLEY
Tight-Knit Community of Tinkerers Keeps Hearts Ticking
MINNESOTA (1950s)

Monica M. Smith

During the 1950s, the Twin Cities of Minneapolis and Saint Paul, Minnesota, began to earn their reputation as an important medical device industry center, later dubbed "Medical Alley." New medical technologies, materials, and procedures developed during World War II, combined with major federal government funding for medical research during and after the war, led to rapid and innovative research and experimentation. The University of Minnesota, particularly its Variety Club Heart Hospital, was a key hub for invention and innovation thanks to charismatic leadership and a culture of collaboration, problem-solving, and risk-taking (Figure 1). Surgeons, medical residents, and engineers—most of whom were born and educated in Minnesota—worked together to develop inventions and innovative techniques related to cardiac surgery that transformed the medical field. As part of a community of health care organizations such as the Mayo Clinic, medical device companies like Medtronic, and bold investors and skilled workers, Medical Alley grew to become a world leader in medical research and care.

PLACE

A Brief History of the Twin Cities

The Twin Cities grew up by Saint Anthony Falls, the northernmost navigable point of the Mississippi River near its intersection with the Minnesota River. This area was rich in natural resources, from rivers for fishing and transportation to vast timber that provided food and shelter for wildlife (and thus offered good hunting) to rich farm land for cultivating crops. When Europeans arrived in the area during the mid-1600s, primarily two Native American tribes were living in Minnesota: the Ojibwa and Dakota. Settlement grew up by trading posts for Native Americans and European trappers, and then in 1820 the American government began building Fort Snelling, around which the Twin Cities arose.

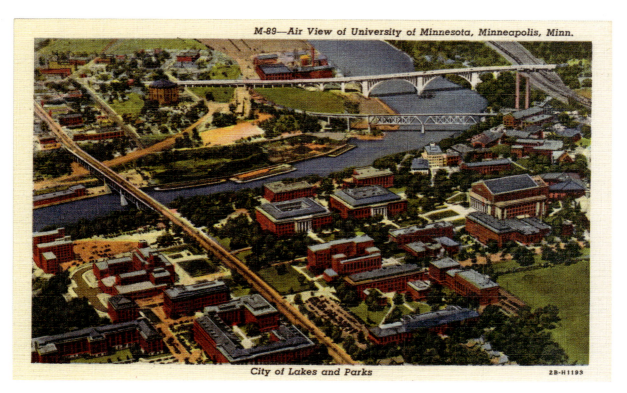

FIGURE 1
Postcard view of the University of Minnesota's Minneapolis campus, 1942. Lake County (IL) Discovery Museum, Curt Teich Postcard Archives.

Established as Minnesota Territory in 1849, Minnesota quickly achieved statehood in 1858. The area's abundant resources, which supported cheap and productive farm land and provided fodder for the growing number of lumber mills and flour mills, attracted the first major wave of German, Norwegian, Swedish, and Irish immigrants during the late nineteenth century. These close-knit groups had similar ethnic backgrounds and religious, educational, and cultural values that arguably helped them work together to build their community in a rather harsh environment with long winters. Once the Great Northern Railway line from Saint Paul to Seattle, Washington, was completed in 1893, the Twin Cities became a major Midwestern hub for transportation of goods and people between the east and west coasts of the United States.[1]

Although Minnesota experienced a second major wave of immigration after the Vietnam War, peaking in the 1980s when refugees from Southeast Asia resettled in the area, the state's population today is only about 7% percent foreign born and is 86% white, with German, Scandinavian, and Irish heritage still predominant.[2] Although the medical industry has attracted more diverse talent from outside the region in recent years, the majority of the inhabitants are still born, raised, and educated in Minnesota, and the state boasts higher than average graduation rates from high school and college.

The Medical Alley story, especially its birth in the 1950s, involves primarily men who grew up in Minnesota; received their undergraduate, graduate, and postgraduate training at the University of Minnesota; lived near each other; and shared similar backgrounds and values. This relative homogeneity, which differs from the other *Places of Invention* stories, seems to have supported and promoted the tight-knit collaboration among the pioneers who helped create Medical Alley.

The University of Minnesota and the Mayo Clinic

Founded in 1851 and established after the Civil War as a land-grant university, the University of Minnesota's first campus in Minneapolis overlooked the Saint Anthony Falls on the Mississippi. Eventually, the university moved about one mile downriver to its current location in Minneapolis and then expanded to campuses in Saint Paul and other locations in the state as it became the region's largest university. In 1887, Minnesota became the first state to require a board examination to receive a medical license, leading to the formation of the university's College of Medicine and Surgery the following year.[3]

Around the same time, English-born Dr. William Mayo brought his sons, Dr. William "Will" Mayo and Dr. Charles Mayo, into his frontier medical practice in Rochester, Minnesota, about 80 miles southeast of the Twin Cities. After a devastating tornado hit Rochester in 1883, the Mayos collaborated with the Sisters of St. Francis to found the first general hospital, Saint Marys, in southeastern Minnesota. With a busy surgical practice at the hospital, the Mayos soon invited other doctors to join their clinic. Dr. Will Mayo declared, "It has become necessary to develop medicine as a cooperative science; the clinician, the specialist, and the laboratory workers uniting for the good of the patient."[4]

The Mayo Clinic became a primary rival of the University of Minnesota for funding, patients, and reputation, but the two institutions also mutually attracted, trained, and shared talented doctors, nurses, and other medical staff. Dr. Charles Mayo even served as a member of the University of Minnesota's board of regents. Together, the two institutions helped the state gain fame for top-notch medical care. During World War I and World War II, the University of Minnesota and Mayo Clinic collaborated to supply and staff Base Hospital Number 26, which was founded at Fort Snelling. In the first war, the unit served in France and cared for nearly 6,000 patients during 1918–19. Once the United States formally entered World War II, students and faculty from the university served at General Hospital Number 26, first in Algeria and then in Italy, where they helped more than 8,000 patients. General Hospital Number 26 became the longest-serving hospital in the Mediterranean Theater.[5]

Medical research changed rapidly thanks to extraordinary wartime advances in medical techniques, tools, and materials. After World War II ended, large amounts of federal government funding continued to support medical research and to encourage veterans' science and technology education through the GI Bill.[6] A shift toward financing and publishing the research conducted by medical faculty and researchers in surgical departments, nursing and dentistry schools, and public health also occurred. In his seminal 1945 report, *Science, the Endless Frontier*, Dr. Vannevar Bush, director of the Office of Scientific Research and Development, argued effectively for postwar federal funding for research and development, including funding for medical research at universities:

> With the advent of the transistor, a variety of plastics, and other new synthetic materials, medical technology was on the threshold of several major breakthroughs that would dramatically transform the diagnosis and care of patients around the world.
>
> **EARL BAKKEN**

> The responsibility for basic research in medicine and the underlying sciences, so essential to progress in the war against disease, falls primarily upon the medical schools and universities. Yet we find that the traditional sources of support for medical research in the medical schools and universities, largely endowment income, foundation grants, and private donations, are diminishing and there is no immediate prospect of a change in this trend. Meanwhile, the cost of medical research has been rising. If we are to maintain the progress in medicine which has marked the last 25 years, the Government should extend financial support to basic medical research in the medical schools and in universities.[7]

The combination of major funding, an educated workforce, and an emphasis on both basic and applied science and technology research spurred further invention and innovation in many industries, including but not limited to the medical field.

Surgical Innovation at the University of Minnesota

From 1930 to 1967, the University of Minnesota's Department of Surgery was chaired by Dr. Owen H. Wangensteen (Figure 2), a revered mentor who famously nurtured research-based experimentation, risk-taking, and creative thinking among the staff. Minnesota native Wangensteen, who earned his B.S., M.D., and Ph.D. from the University of Minnesota and served as a fellow at the

FIGURE 2

Dr. Owen H. Wangensteen, chief of surgery, University of Minnesota, 1966. Image courtesy of University of Minnesota Archives, University of Minnesota–Twin Cities.

Mayo Clinic, became famous for his development of a suction apparatus for bowel procedures nicknamed "Wangensteen alleys," which helped save thousands of lives during surgery on the World War II battlefront. Ogden Nash even memorialized Wangensteen's invention in a poem on a card sent to the doctor and his staff in 1951:

> May I find my final rest in
> Owen Wangensteen's intestine
> knowing that his masterly suction
> will assure my resurrection.[8]

Wangensteen's emphasis on research proved to be financially astute, as it strongly positioned the university's Department of Surgery to receive government funding available after World War II for pursuing cutting-edge medical research. Dr. C. Walton Lillehei recalled,

> [Wangensteen] had a great faith in research, animal or other types of laboratory research. He felt that the results of his research gave the young investigator courage to challenge accepted beliefs and go forward, which you would not have had, as I look back, as a young surgical resident. That's why many of the great universities didn't produce much in the way of innovative research, because they were so steeped in tradition. Wangensteen had a wide-open mind. If research showed some value, then you should pursue it.[9]

Called "the Chief" by his residents, Wangensteen was first and foremost a mentor with an eye for talent. "If he was convinced that a candidate had a good brain and a capacity for work, he would say on the spot: 'When can you start?'"[10] Wangensteen built up one of the country's greatest surgical departments that became world famous particularly for its innovative cardiac surgical techniques and technologies. His staff who had earned degrees at the University of Minnesota included such pioneers as Minnesota native Dr. Lillehei, often called "the father of open-heart surgery"; Dr. Christiaan Barnard, the South African surgeon who performed the world's first heart transplant; and Dr. Norman Shumway, who became known as the "father of transplant surgery." During Wangensteen's tenure, nearly 1,000 surgeons went through the Department of Surgery, and hundreds ended up serving on surgical faculties across the world.[11]

In 1951, the University of Minnesota opened its Variety Club Heart Hospital, the first hospital in the United States to focus entirely on cardiac patients (Figure 3). The idea for the hospital was initially promoted by Al W. Steffes, who owned movie theaters in Minneapolis and was an office holder in the Variety Club, a service organization of people in the entertainment business. He worked with his friend Dr. Morse Shapiro at the University of Minnesota to make the case to his

FIGURE 3
University of Minnesota Variety Club Heart Hospital, 1955. Image courtesy of
University of Minnesota Archives, University of Minnesota–Twin Cities.

local chapter of the Variety Club to donate funds for a specialized heart hospital to serve the needs of children throughout the upper Midwest. In the end, the Variety Club raised about one-third of the hospital's $1.5 million project funds, in part through the use of a short fundraising film starring Hollywood actor Ronald Reagan that showed as a trailer at movie theaters across the region. The federal government and the university paid the remaining two-thirds of the project budget.[12]

PEOPLE

Dr. C. Walton Lillehei

The new Variety Club Heart Hospital gained rapid fame when Wangensteen's protégés Dr. F. John Lewis, Dr. Richard Varco, and Dr. C. Walton "Walt" Lillehei completed the world's first open-heart operation to repair the heart of a five-year-old patient on 2 September 1952. Lillehei was among the university's surgical residents who had gained valuable experience working in General Hospital Number 26 as a member of the U.S. Army Medical Corps during World War II, earning a Bronze Star for his service and rising to the rank of lieutenant colonel. His wartime experience as a surgeon helped him build experience and confidence in taking risks, and his subsequent training as

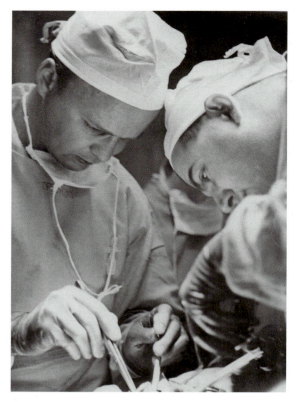

FIGURE 4

Dr. C. Walton Lillehei (left) and Dr. Richard Varco (right) in surgery at the University of Minnesota Variety Club Heart Hospital, 1954. Image courtesy of University of Minnesota Archives, University of Minnesota–Twin Cities.

a resident under Wangensteen encouraged risk-taking (Figure 4).

Developing safe, reliable cardiac surgery was a major medical quest by the 1950s. One of the challenges was figuring out how to have the patient's damaged heart stay dry during surgery yet keep oxygenated blood properly circulating through the body. Lillehei had been developing techniques to help "blue babies," or children who could not get adequate oxygen because they were born with a heart defect. They had very short life expectancies without surgery. Initially, open-heart surgical procedures, starting with the famous operation in 1952, used hypothermia—cooling down the patient's body temperature—to slow the heartbeat. However, that technique did not allow for enough time to conduct complex surgery.

In 1954, Lillehei decided to attempt open-heart surgery using an incredibly risky and controversial cross-circulation technique. Developed especially for young children unable to handle hypothermia, cross circulation involved stopping the patient's heart during surgery and linking the patient's circulation system to that of a donor, such as a parent. Considering the 200% mortality risk of two lives to try to save one, most surgeons and University of Minnesota Medical School faculty and staff opposed it. However, Wangensteen supported experimentation and risk-taking. When he was notified that an acceptable donor-patient duo had been found, he wrote Lillehei a simple note: "Dear Walt, by all means, go ahead! Good luck! O.H.W."[13] On 26 March 1954, Lillehei and his team successfully conducted the first open-heart procedure using cross circulation, which shocked and amazed the medical field (Figure 5).

Katherine Lillehei, who earned a nursing degree at the University of Minnesota and met her husband Walt while working as a nurse at the hospital, recalled that these were very exciting, although stressful, times for him and his colleagues. "We moved to be close to the hospital. There was one thing that my husband told [our sons], 'When you're looking for a house, be close to the hospital if you're a surgeon.'"[14] Walt would bring fellow surgeons over to the Lillehei house after an operation to talk and decompress. The surgical team members worked long hours and were

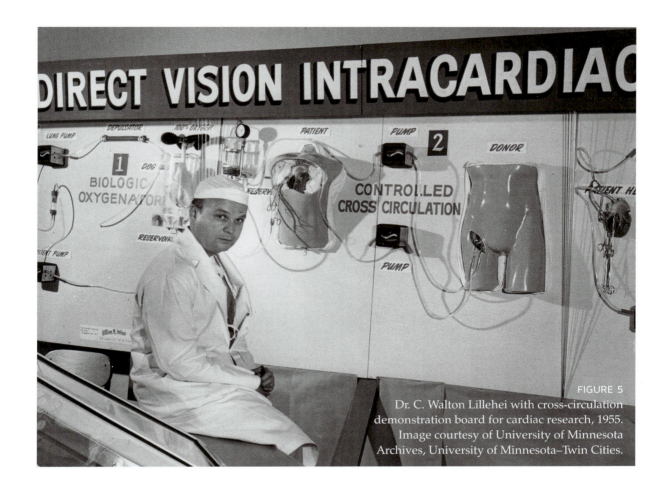

rarely home with their families, let alone out socializing, although the Lilleheis liked to host parties when they could. Most of their relationships revolved around the hospital.

For decades surgeons and engineers had attempted different ways to develop a mechanical heart-lung machine that would provide a safer way to keep the patient's blood oxygenated and circulating during cardiac surgery. Dr. John Gibbon at Philadelphia's Jefferson Medical College, with IBM, developed a heart-lung machine featuring three roller pumps and a stationary screen oxygenator. The first operation using Gibbon's Model II oxygenator in 1952 was on a fifteen-month-old girl, who died of unrelated causes; the second surgery in 1953 was on a college student, who survived. Two more operations ended in deaths, and Gibbon decided it was too risky to continue.

In 1955 surgeons at both the Mayo Clinic and the University of Minnesota tried to improve upon Gibbon's work. Dr. John Kirklin collaborated with Mayo Clinic colleagues to build a heart-lung machine based on Gibbon's blueprints using DeBakey pumps and stainless-steel screens. The resulting Mayo-Gibbon machine worked, but it was very expensive, needed to have lots of blood pumped through it, and was complicated to disassemble, sterilize, and reassemble between surgeries (Figure 6).

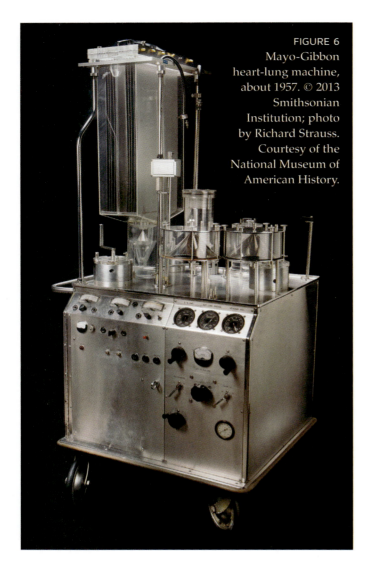

At the same time Lillehei asked his surgical resident Dr. Richard DeWall to create a simpler, much less expensive blood oxygenator. DeWall was a recent graduate of the University of Minnesota Medical School who had started working in Lillehei's lab in 1954. "After about a dozen cross-clinical procedures, Dr. Lillehei suggested to Dr. DeWall that it would be a good research project to develop an oxygenator to avoid the need for a donor. Doctor DeWall accepted the challenge."[15] Within just a few months they had assembled a surprisingly simple, single-helix "bubble" oxygenator made out of cheap polyethylene tubing from Mayon Plastics in Minneapolis, which supplied this tubing to dairies for separating milk and cream.[16]

When the helical tube reservoir was filled with blood, any bubble-containing blood became trapped at the top and normally oxygenated blood could be retrieved from the bottom of the coiled reservoir. After the device was tested on about 70 dogs, the DeWall-Lillehei bubble oxygenator was ready for human use in May 1955. DeWall continued to improve the invention through three models, but it remained a simple, inexpensive, and easily sterilizable device that could be built to accommodate the amount of blood required for each patient. DeWall and Lillehei's patented helical oxygenator made open-heart surgery a much more practical procedure (Figures 7–9).

Another one of Lillehei's residents, Dr. Vincent Gott, improved upon DeWall's design by inventing a bubble oxygenator that flattened and enclosed the helix between two heat-sealed plastic sheets. This sheet bubble oxygenator, developed in 1957, proved to be a key factor for widespread acceptance of open-heart surgery because it was easy to manufacture and distribute and it was inexpensive enough to be disposable.[17] Soon open-heart surgery spread beyond the two pioneering institutions, the University of Minnesota and the Mayo Clinic.

FIGURE 7
DeWall-Lillehei bubble oxygenator, or heart-lung machine, 1955. © 2014 Smithsonian Institution; photo by Richard Strauss. Courtesy of the National Museum of American History.

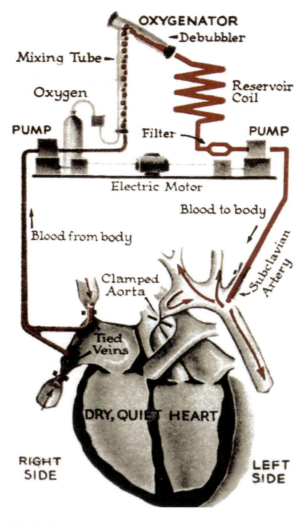

FIGURE 8
Schematic diagram of the helical reservoir oxygenator published by coinventor Richard DeWall in "The Evolution of the Helical Reservoir Pump-Oxygenator System at the University of Minnesota," *Annals of Thoracic Surgery* 76, no. 6 (December 2003): S3320–S2215. Courtesy of Richard A. DeWall, M.D.

As evidenced in other stories featured in this book, charismatic leaders—in this case Lillehei and his boss Wangensteen—often play key roles in attracting and nurturing talent to help create a place of invention. Medical students and residents plus surgeons from around the United States and from foreign countries were attracted to studying at the University of Minnesota because of Lillehei's fame for his revolutionary open-heart surgery techniques and technologies (Figure 10). He proved to be an influential mentor for many cardiac surgeons. According to Gott, 134 cardiothoracic surgeons were trained directly by Lillehei at the University of Minnesota between

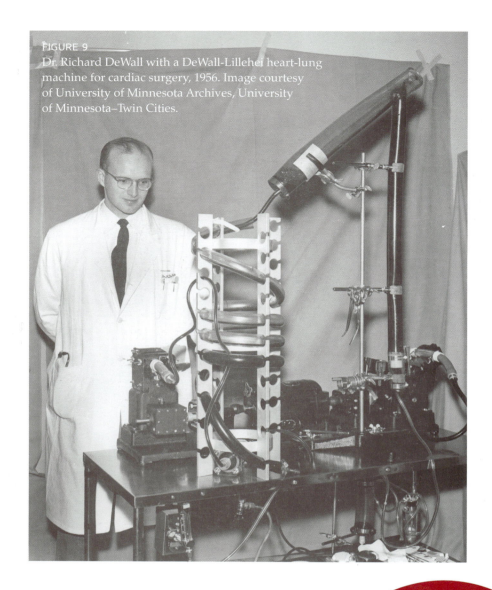

1951 and 1967, plus another 20 at the New York Hospital–Cornell Medical Center where Lillehei worked from 1967 to 1974.[18] Hundreds more visited him, watched him in surgery, read his publications, or were otherwise influenced by his pioneering work. "As one of cardiac surgery's most productive innovators, he will be remembered for his ingenuity, imagination, and boldness."[19]

Earl Bakken

"Medtronic co-founder Earl Bakken was fascinated by electricity from an early age. Even as a toddler, he played with wires,

> We had many visitors at the time from all over the world, because the only place that was doing open-heart surgery on a regular basis from 1952 through 1955 was the University here in Minneapolis.
>
> DR. C. WALTON LILLEHEI

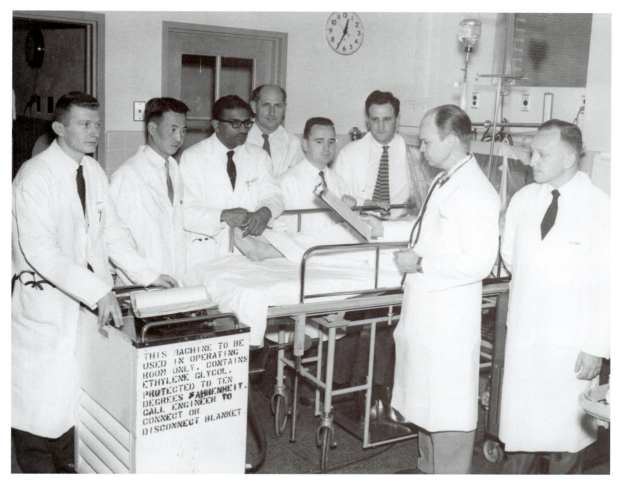

FIGURE 10
Dr. C. Walton Lillehei with foreign residents at the University of Minnesota in 1958. Image courtesy of University of Minnesota Archives, University of Minnesota–Twin Cities.

cords, plugs, and connectors—leading an uncle to warn his encouraging mother, 'That boy's going to electrocute himself someday.' But his mother kept Earl supplied with spare parts, and he spent much of his . . . childhood wiring up his own creations."[20] Bakken credits seeing the 1931 *Franken-stein* movie starring Boris Karloff at the Heights Theater in Minneapolis when he was about eight or nine years old as the pivotal moment when he was inspired to become an electrical engineer (Figure 11). He recalled,

> What intrigued me the most, as I sat through the movie again and again, was not the monster's raving, but the creative spark of Dr. Frankenstein's electricity. Through the power of his wildly flashing laboratory apparatus, the doctor restored life to the unliving.[21]

FIGURE 11

A special screening of *Frankenstein* was held at the historic Heights Theater in Minneapolis on 2 December 2008 to celebrate Earl Bakken's 85th birthday. Courtesy of Medtronic, Inc.

He continued playing with electricity at home and school, becoming the go-to kid for operating and maintaining the public address system, movie projector, and other electrical equipment.

During World War II, Bakken served as an airborne-radar maintenance instructor in Florida and then took advantage of the GI Bill to earn his B.S. degree in electrical engineering at the University of Minnesota's Institute of Technology. He started to pursue a master's degree at the institute but found himself more interested in "wander[ing] across Washington Avenue to the University Hospitals, where I became acquainted with some of the people in their extensive labs and began providing, at their request, an ad-hoc, on-the-spot repair service for malfunctioning equipment."[22]

At a party, Palmer Hermundslie suggested to his friend Bakken, who was essentially also his brother-in-law (the two men had married sisters Elaine and Connie Olson, respectively), that they should set up a business to service electrical medical equipment, with Hermundslie as manager and Bakken as the technician. Bakken said,

> After World War II, hospitals were just starting to use electronics. Since there weren't any electronic repair businesses, they would send their equipment to radio

shops for repair that were not equipped to do a good job. So we thought there might be a business opportunity. Seemed like a good time.[23]

Their company, Medtronic, Inc., had humble beginnings. Like more famous garages in Silicon Valley that begat Apple and Hewlett-Packard, the Hermundslie family's 800-square-foot garage (made out of two railway boxcars) served as a convenient location for Bakken and Hermundslie to found their company in April 1949 (Figure 12). Hermundslie described it as follows:

> My dad and I had bought the boxcars from Great Northern and disassembled them on the Columbia Heights rail yards. Before Medtronic moved in, we had used the garage to rebuild wrecked cars when I was in high school. Earl and I spent most of that first month just cleaning up the building. It was still pretty crude—no plumbing or ceiling and no window casings.[24]

They earned minimal revenue during their company's first year, including just $8 in the first month. To stay in business, Palmer continued working a day job at a lumberyard, Earl repaired television sets, and Earl's wife Connie supported their family through her job as a medical technician

FIGURE 12
The Hermundslie family garage around 1930, where Palmer Hermundslie and Earl Bakken cofounded Medtronic, Inc., in 1949. Courtesy of Medtronic, Inc.

at Northwestern Hospital. (Both she and her sister Elaine earned medical-related degrees from the University of Minnesota.) Fortunately, Medtronic's prospects improved when they began not only repairing equipment but also selling medical equipment for the Sanborn Company, which meant traveling around the region and working with many doctors, nurses, and technicians, including staff at the University of Minnesota. Bakken reflected,

> I was an outsider—a non-physician—in their highly specialized and incredibly demanding world. But I think my independence was a quality they appreciated. If they had a problem with one of their electrical devices, I could often fix it immediately. . . . Furthermore, I was about the same age as most of the interns and residents I'd met there, so we shared a lot of life experiences and attitudes. . . . Later on, as it happened, many of those young doctors became prominent physicians and heads of surgery at hospitals around the world, thus giving Medtronic far-flung and influential contacts when we were in the pacemaker business.[25]

Most notably, his acquaintance with Lillehei at the university turned into an important and mutually beneficial partnership. Lillehei later joked, "We both are of Norwegian descent and the name Lillehei in Norway means 'small hill.' Bakken in Norway means a 'small hill' or a 'rounded hill.' So we were destined to work together, I guess."[26]

INVENTION
Lillehei, Bakken, and the Early Days of the Pacemaker Revolution

On Halloween night 1957, a sudden Twin Cities–wide blackout helped spawn Lillehei and Bakken's collaboration that would end up transforming the medical field. At that point in time, Lillehei's young open-heart patients would be hooked up for several days during postsurgery recovery to large, bulky, AC-powered pacemaker machines, which had to be moved around on wheeled carts and kept plugged into wall outlets. "It turned out that the surgery actually damaged the heart's natural conduction system, and they needed these external pacemakers, which used electrodes on the chest, to get the children's hearts beating at the proper rate and give them time to recover from the surgery."[27] Lillehei had asked a medical technician to try to invent a more portable battery-powered pacemaker, but nothing had been produced. When the blackout occurred and backup generators failed, patients' lives were put in jeopardy, spurring Lillehei to act.

The day after Halloween, Lillehei spied Bakken in the hallway and called after him, "Hey, Earl, I've got a problem for you."[28] He knew that Bakken had the necessary knowledge and skills because of the work Bakken had done repairing machines at the hospital, and Lillehei also admired the engineer's interest in repeatedly watching him perform open-heart surgery. Lillehei noted,

"Electricians absolutely refused to come into the operating room when an operation was in process. Earl Bakken came over for every open-heart operation."[29]

Bakken accepted the challenge without hesitation. Initially, he "was just going to use a car battery and an inverter and a recharger, which was probably about a hundred pounds of apparatus."[30] However, while working in Medtronic's garage office, Bakken found inspiration in a surprising source: *Popular Electronics*. Its April 1956 issue had an article titled "Five New Jobs for Two Transistors" that included a short column about a metronome circuit using transistors (still a relatively new invention).[31] Bakken said,

> One of the home projects in there was to build a little box that would put clicks on a loudspeaker, the way a metronome sets clicks for timing music. It's interesting that a metronome has the same rate range as a normal heart. You can adjust it from 50 up to 100 or 150 or so pulses per minute. So I plagiarized that circuit, in effect, [and] modified it a bit so it would be a one-millisecond pulse.[32]

Bakken drew schematics on envelopes and grocery bags and developed a prototype external wearable cardiac pacemaker in an unbelievably short period of four weeks (Figures 13 and 14).

On display now at The Bakken Museum in Minneapolis, this pacemaker prototype is basically just a small aluminum box containing two transistors and a nine-volt photoflash battery, with a neon bulb to show it pulsing (Figure 15). The device was intended only for animal testing—Lillehei

FIGURE 13
Earl Bakken at work in the Medtronic garage in Minneapolis, about 1955. Courtesy of Medtronic, Inc.

FIGURE 14
Employees Roger Hanson (left) and Bob Wingrove (right) in the Medtronic garage, about 1955. Courtesy of Medtronic, Inc.

dubbed it the "dog model" since it was tried first on a dog—but Lillehei put it to use almost immediately. Bakken was stunned to return to the hospital and see it attached to a young cardiac patient within days of its invention.

Although it functioned effectively, they quickly realized the device's exterior design needed tweaking. As Bakken said,

> There couldn't be a better example of technology transfer and economic development than the University of Minnesota-Medtronic story. It turned out to be the beginning of Medical Alley, with all the medical companies that were started later by former Medtronic employees.
>
> DR. C. WALTON LILLEHEI

> This did the job of pacing the child's heart, keeping him alive until they started conducting on their own. But the children would start playing with the knobs, turning it on and off, and, of course, the nurses had to tape all this up when they used them. So we had to . . . put the pacemaker into a box where the little fingers couldn't get in there and change the knob and with its on/off switch that was difficult to use.[33]

He worked with his employees to develop a black plastic-covered version of the pacemaker. About ten of these went into clinical use at the University of Minnesota, but Medtronic decided black was

FIGURE 15

The original prototype for the battery-operated, transistorized, wearable cardiac pacemaker invented by Earl Bakken in 1957. Library and Artifact Collections of The Bakken Museum, Minneapolis.

FIGURE 16

The Medtronic 5800 external cardiac pacemaker model, first produced commercially in 1958. This object in the National Museum of American History's collection dates from about 1972. © 2014 Smithsonian Institution; photo by Richard Strauss. Courtesy of the National Museum of American History; gift of Medtronic, Inc.

not an appropriate color for hospitals. Later in 1958, the company began manufacturing a commercial version in white plastic that was called the 5800 in honor of its birth year (Figure 16). The three versions were essentially identical in circuitry and other interior features.[34] Bakken believed "it was probably one of the first really commercial uses of transistors. Of course, operating at such low currents and low voltage . . . we felt they were very safe."[35]

Weighing about 10 ounces, the Medtronic 5800 pacemaker connected to the patient's heart through two wires inserted into the skin and could be worn in a sling or, according to the company, "a special pocket designed for women that could be attached to a bra" (Figure 17).[36] It was intended only for short-term use while patients recovered from surgery. However, in 1959, Warren Mauston became the first patient to use it as a long-term pacemaker, thanks to the new Hunter-Roth bipolar platform electrode developed by Dr. Samuel Hunter (who had coincidentally done a surgical residency under Walt Lillehei in 1956–57) and Medtronic engineer Norman Roth.[37] Mauston, a seventy-two-year-old businessman from Saint Paul, survived for six years wearing the 5800 and proudly helped advertise the pacemaker in articles and a 1959 Medtronic promotional video. As quoted in the *Saturday Evening Post*, Mauston boasted, "Considering what might have happened without this [pacemaker], I'm the luckiest man in ten states."[38]

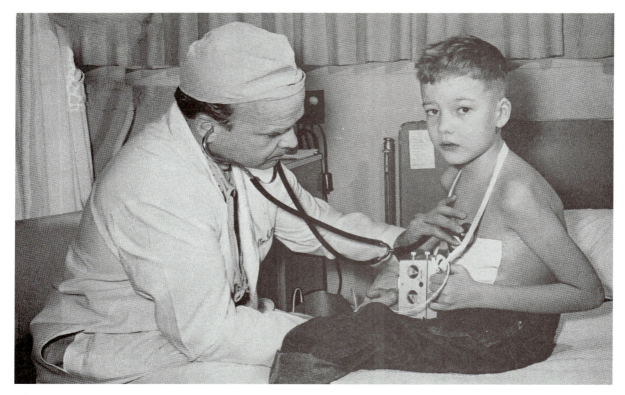

FIGURE 17
Dr. C. Walton Lillehei with patient David Williams, who is wearing the Medtronic 5800 cardiac pacemaker. Featured in the *Saturday Evening Post*, 4 March 1961. Photograph © SEPS licensed by Curtis Licensing, Indianapolis, IN. All rights reserved.

Medtronic quickly gained publicity and fame as "the general press picked up the story of the little metal box that kept children's hearts beating, and people were calling it a 'miracle.'"[39] Anecdotes swirled about patients whose lives were saved and revitalized thanks to the 5800. Bakken recalled, "We had this veteran from the VA Hospital out here who would go be released each weekend and he'd go up to northern Minnesota where they went square-dancing, and he would invariably break his [pacemaker] wire, and then I'd get called to solder it together on Monday morning."[40] Medtronic and other companies in the medical device field knew they needed to develop a fully implantable cardiac pacemaker for safer, more practical long-term use. The race was on, and several teams around the world were working on the problem.[41]

Wilson Greatbatch, an electrical engineer working in his barn workshop, collaborated with Dr. William Chardack and Dr. Andrew Gage, surgeons at the Veterans Administration Hospital in Buffalo, New York, to develop and successfully implant one of the first devices in April 1960. Greatbatch said later, "I seriously doubt if anything I ever do will give me the elation I felt that day when a two-cubic-inch electronic device of my own design controlled a living heart."[42] Greatbatch and Chardack's work first came to Medtronic's attention when they contacted the company to

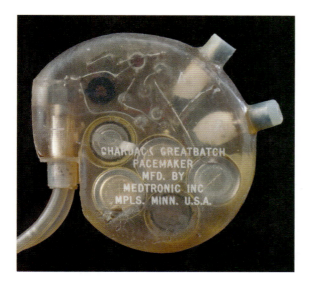

FIGURE 18
Chardack-Greatbatch implantable cardiac acemaker model manufactured by Medtronic, Inc., about 1961. © 2014 Smithsonian Institution; photo by Richard Strauss. Courtesy of the National Museum of American History.

learn more about the Hunter-Roth electrode being used on pacemakers. Medtronic immediately recognized the importance of the fully implantable Chardack-Greatbatch pacemaker (Figure 18). Palmer Hermundslie flew his own airplane out to Buffalo in October 1960 to negotiate a contract to license it for ten years, thus grabbing the opportunity to develop and manufacture the technology in Minnesota.

Medtronic's business grew initially thanks to the success of the 5800 model and then to manufacturing and selling the Chardack-Greatbatch implantable pacemaker, which allowed them to start building new 15,000-square-foot headquarters in St. Anthony Village, a Minneapolis suburb. The garage era was over (Figure 19). As Bakken recalled,

FIGURE 19
Medtronic cofounders Earl Bakken (far left) and Palmer Hermundslie (far right) with the "Garage Gang" in February 1959. Courtesy of Medtronic, Inc.

The contrast between our humble beginnings back there off Central Avenue and the new operation [which opened in spring 1961] seemed almost comical. The new place even included a cafeteria, library, and auditorium—a far cry, to be sure, from the cramped work space and homemade desks of the garage! At the same time, it seemed inconceivable that we'd have to build two additions comprising 75,000 square feet before the end of the new decade.[43]

THE RISE OF MEDICAL ALLEY

The origins of Medical Alley mostly involved Minnesota-born doctors and engineers collaborating on inventive medical techniques and technologies related to cardiac surgery at the University of Minnesota and Medtronic. The main rival to the university was the Mayo Clinic, which continues to be highly regarded for its health care services. Other players in the medical technology and materials business in the early days included 3M, arguably the most famous company in Minnesota. Together, these companies and institutions started to earn the state its reputation as a major hot spot for medical devices, materials, and health care.

Medtronic's expansion during the 1960s and 1970s benefitted from financial leadership by Palmer Hermundslie and then other investors and board members, from the license agreement and resulting friendship between Earl Bakken and Wilson Greatbatch, from Walt Lillehei's referrals, and also from Bakken's strong belief in what he called the "high-tech, high-touch" approach of cutting-edge engineering combined with a concern for human wellness.[44] Medtronic had some business struggles during the 1960s, but it went on to become a multibillion-dollar company with dozens of spin-offs.

As in Hartford, Connecticut, Silicon Valley, and Fort Collins, Colorado, skilled workers and risk-taking entrepreneurs left companies like Medtronic to find new jobs or create their own start-ups, to improve existing products or develop new ones, and to provide the labor, materials, tools, and other supplies and services needed for the local industry. Over time, this growing network of participants in and suppliers for the medical device industry created the region later dubbed Medical Alley, which ranged from Rochester through the Twin Cities up to Duluth on the shores of Lake Superior. The area's population has diversified since its beginnings, but compared to many places that relied on immigration from other regions and countries, Medical Alley in general retained a relatively homogeneous, Minnesota-born and -educated workforce. Inventors and innovators reflected

> Why Minnesota? Where else are there 10,000 engineers all in one place with such medical expertise?! It's a highly skilled, tight-knit, hard-working community and I would not consider founding my companies anywhere else.
> **MANNY VILLAFAÑA**

FIGURE 20

This Microlith P 0505 pacemaker in the National Museum of American History's collection was manufactured in 1978 by Cardiac Pacemakers, Inc., in Minneapolis. © 2014 Smithsonian Institution; photo by Richard Strauss. Courtesy of the National Museum of American History.

primarily the community of German, Scandinavian, and Irish descendants, who often were from Lutheran or Catholic religious backgrounds and were educated at the University of Minnesota.

One notable outlier is Manuel "Manny" Villafaña, who grew up in the South Bronx, New York, with Puerto Rican parents. A high school graduate, he quickly showed his skills as a salesman. By his early twenties, Villafaña worked for Picker International selling medical products on behalf of many companies, including Medtronic. In 1967 Earl Bakken and his colleague Charlie Cuddihy met with Villafaña in New York and recruited him to help expand international distribution of Medtronic cardiac pacemakers. After five years Villafaña left to start rival Cardiac Pacemakers, Inc. (CPI), with three other former Medtronic employees, and they developed a pacemaker using a new lithium battery developed by Wilson Greatbatch that Medtronic had rejected (Figure 20). At CPI Walt Lillehei's brother Dr. Richard Lillehei, who was also a pioneering surgeon at the University of Minnesota, served as a key advisor.

In 1976 Villafaña decided to move on again to start another company, St. Jude Medical. This time his company focused on developing a mechanical heart valve, which became the industry's standard, and he brought on Dr. Walt Lillehei as the company's medical director. Both CPI (now a subsidiary of Boston Scientific Corporation) and St. Jude Medical became Medtronic's major competitors. Since then, Villafaña has founded several more companies in the Twin Cities, most recently Kips Bay Medical, which is named after the New York City Boys and Girls Club where he spent a lot of time as a kid and which he credits in part for his later success (Figure 21).

During the 1970s and 1980s, medical technology companies in Minnesota benefitted financially in part from a new influx of local venture capital that previously targeted the state's mainframe computer industry. Starting in the late 1940s with pioneer William Norris and the company he cofounded, Engineering Research Associates, Minnesota grew as a hub for the burgeoning computer business. In 1957 Norris started a new company, Control Data Corporation, which became one of the world's largest mainframe computer companies. Seymour Cray, dubbed "the father of super-computing," trained as an engineering student at the University of Minnesota, worked with Norris at Control Data Corporation, and eventually founded his own company, Cray Research, Inc., with business headquarters in Minneapolis.[45] Another mainframe computer giant, Honeywell, had its headquarters in the state, too. When the mainframe computer industry collapsed in the 1980s

FIGURE 21
Manny Villafaña (rear) with Dr. Nazih Zuhdi (left), Dr. C. Walton Lillehei (center), and Dr. Christiaan Barnard (right) in Minneapolis on 10 May 1985. Courtesy of Manny Villafaña.

because of the rapid rise of personal computers coming out of Silicon Valley, Minnesota venture capitalists shifted their attention to funding local companies involved in medical invention and innovation.[46]

In 1984, a 501(c) nonprofit trade association called Medical Alley was established to support Minnesota's medical device and health care industry. On 15 September that same year, Rochester's *Post-Bulletin* featured the headline "Hi-Tech Medical Alley's Pulse Is Strong," the first in a week-long series about the state's "mushrooming 300-plus medical high-tech industries. . . . More than 50 percent of the firms are clustered in the Minneapolis-St. Paul metropolitan area, where venture capital and innovative biomedical scientists are available in abundance."[47] The trajectory continued upward, and in 1992 *Businessweek* highlighted Medical Alley as one of the country's high-tech "hot spots."[48]

In 2005 LifeScience Alley was created when Medical Alley and the state's biotechnology association, MNBIO, merged. Today, "nearly 700 member organizations come from all sectors of the life sciences ecosystem, representing medical technology and equipment manufacturers, pharmaceutical and biopharmaceutical companies, health care providers and insurers, agricultural and industrial biotechnology organizations, academic institutions and government entities, and a broad range of service and consulting companies."[49]

Medical Alley has come a long way since the 1950s when Dr. C. Walton Lillehei, with fellow surgeons at the University of Minnesota, performed the first open-heart procedures and Earl Bakken invented the wearable pacemaker to keep Lillehei's patients' hearts beating regularly during recovery. These Minnesota natives, who leveraged each other's inventive skills to revolutionize the cardiac field, were members of a larger ecosystem of charismatic leaders, medical staff, engineers, scientists, investors, suppliers, and others who shared the values of collaboration, adaptability, problem-solving, and risk-taking. The early days of Medical Alley depended primarily on a locally born workforce; more recently, this medical community has benefited greatly from "the importing and aggregation of talent and technology to the cluster."[50] Although not nearly as famous as California's Silicon Valley, Minnesota's Medical Alley is an important place of invention that continues to flourish as one of the world's major medical device industry and health care centers today (Figure 22).

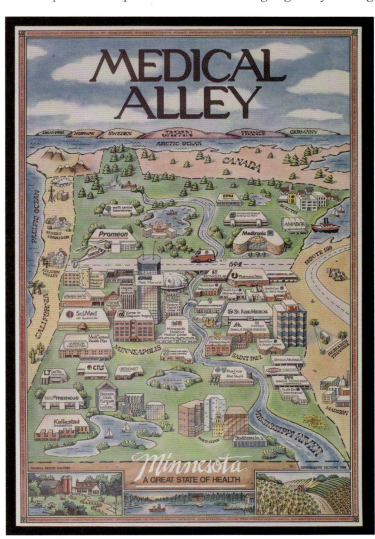

FIGURE 22

"Medical Alley, Minnesota: A Great State of Health" poster created by Synergistic Designs in 1988. Note the Mayo Clinic (lower center); 3M, St. Jude Medical, and the University of Minnesota (center right); and Medtronic (upper right). Also, note that Silicon Valley is highlighted in California on the left, and Route 128 in Massachusetts, the Research Triangle in North Carolina, and Madison, Wisconsin, are all depicted on the right. © 2013 Smithsonian Institution; photo by Jaclyn Nash. Courtesy of the National Museum of American History.

HARTFORD

Factory Town Puts the Pieces Together in Explosive New Ways
CONNECTICUT (Late 1800s)

Eric S. Hintz

In 1868, during his travels "in the East," American humorist Mark Twain provided a San Francisco newspaper with his impressions of one of America's leading cities: "I am in Hartford, Connecticut, now. . . . I think this is the best built and handsomest town I have ever seen." Twain called Hartford "the centre of Connecticut wealth" and described the city's insurance companies and subscription publishing houses, the Sharps rifle factory, and, "last and greatest," the "Colt's revolver manufactory."[1] Twain was impressed, and just six years later, he moved his family into a stately red-brick mansion on Hartford's Farmington Avenue, where he would author several of his classics, including *The Adventures of Tom Sawyer*, *The Adventures of Huckleberry Finn*, and *A Connecticut Yankee in King Arthur's Court*.

In the late 1800s, Hartford was indeed one of America's key places of invention and a leading industrial city (Figure 1). Founded in 1636, the state capital had long been a trading post for merchants shipping goods along the Connecticut River to New York City and the Atlantic. In the 1850s and 1860s, firms like Aetna and Travelers emerged to underwrite the maritime trade, making Hartford the Insurance Capital of the World. Meanwhile, the Colt Armory and its neighboring firms perfected the techniques of interchangeable parts manufacturing, establishing Hartford as one of the birthplaces of American mass production. Thus, all kinds of products, including firearms, sewing machines, typewriters, bicycles, and automobiles, were manufactured in Hartford, making this New England city a hotbed of "Yankee ingenuity" from the late 1800s through the early twentieth century. Unfortunately, Hartford's fortunes changed greatly after World War II, as the effects of deindustrialization turned the city into one of the poorest in the nation. However, Hartford's leaders have drawn inspiration from the city's innovative past to revitalize the city and muster a comeback.

FIGURE 1

City of Hartford, Conn., 1864. In this bird's-eye view of the city, lithographer John Bachmann captured the factories and smokestacks of Hartford, as the observer looks east toward the Connecticut River. The Connecticut Historical Society.

PLACE

The Connecticut River Valley Inspires Yankee Ingenuity

Hartford owes much of its economic and industrial success to its geological and natural endowments, especially the Connecticut River. Millions of years ago, tectonic forces created a long, narrow rift valley in present-day New England. Over the millennia, erosion carved out the Connecticut

River, which today runs 410 miles south from the Canadian border through Vermont, New Hampshire, Massachusetts, and Connecticut before emptying into the Long Island Sound at Saybrook. The result was the alluvial Connecticut River valley, which has less rugged terrain, more fertile soil, a marginally warmer microclimate, and a slightly longer growing season than the rest of New England.[2]

Thus, when Adriaen Block of the Dutch West India Company became the first European to sail up the Connecticut River in 1614, he reported the native populations living prosperously on crops such as maize, squash, and beans grown in the valley's fertile soil. The Dutch established a small fort and fur trading post called Huys de Goede Hoop along the Connecticut River at present-day Hartford. Successive smallpox epidemics in 1616–19 and 1633–34 wiped out much of the native population, clearing the way for additional European settlement. Led by Reverend Thomas Hooker, English settlers moved west from the Massachusetts Bay Colony in 1633–34 and drove the Dutch from their fort, renaming the area Hartford, after Hertford, England.[3] The Dutch derisively referred to the invading Englishmen as *Jankes* (literally, "little Johns"); the Anglicized term *Yankee* eventually came to describe people and things native to New England, particularly Connecticut.[4]

The Connecticut River and its surrounding hinterland provided the foundation for Hartford's industrial prosperity.[5] The river is navigable by deep-draft sailing vessels as far north as Hartford and provided a gateway to New York City and the Atlantic. Thus, Hartford became a natural trading post and center of mercantile exchange. Like the natives before them, the English settlers prospered agriculturally in the fertile soils of the Connecticut River valley and sold their surpluses to merchants who set up wharves at the river's edge. Exports often sailed for the Caribbean; cargoes included cattle and cured beef, horses and mules, lumber, barrel staves, coarse cloth, hardware, dried fish, live poultry, pearl and potash, salt pork, tobacco, onions, and dried corn. Hartford imported rum, sugar, molasses, coffee, and spices from the West Indies and, with American industries still in their infancy, finished goods from England.[6]

> Get a bicycle. You will not regret it, if you live.
>
> MARK TWAIN, "TAMING THE BICYCLE"

The international shipping business had many inherent risks, including cargo losses from sunken ships and piracy, plus the specter of fire and theft at the merchants' riverside warehouses. To protect against these losses, Hartford's merchants, led by Jeremiah Wadsworth, entered into informal, mutual partnerships to pool their risks, and in 1794 Hartford's insurance industry was born. Several new insurance companies emerged in the early nineteenth century, including Hartford Fire Insurance ("the Hartford," 1810), Aetna (1819), and Travelers (1864). After 150 years, many of these firms are still in business, making Hartford the Insurance Capital of the World. Meanwhile, the Hartford Bank (founded 1792) and Society for Savings (1819) grew to serve the financial dealings of the sailors, merchants, and insurers. Thus, by the mid-nineteenth century, Hartford had become one of the nation's premier insurance and financial centers.[7]

Meanwhile, the steam power revolution had a tremendous impact on Hartford trade and industry. In 1815, the *Fulton* became the first steamboat to chug up the Connecticut River; thereafter, regular steam service increased trade by quickening the 40 mile journey upriver from the Long Island Sound. The first steam railroad arrived in Hartford in 1839, taking advantage of its strategic location midway between New York City and Boston and interconnections with the river trade. Meanwhile, steam engines transformed Hartford into a factory town. Although the earliest American mills grew up next to rivers, Hartford suffered from "flat water"—that is, the Connecticut River did not fall enough in elevation to sufficiently power a water wheel. However, steam severed the ties between geography, topography, and industrial power, and for manufacturers, it made good economic sense to operate in a densely populated transportation hub like Hartford, which featured a large local workforce and the steamship and railroad connections to bring raw materials in and ship finished goods out.[8]

Mercantile profits provided the venture capital for Hartford's manufacturing industries. According to Hartford historian Ellsworth Grant, the maritime trade gave rise to "that class of able, daring, successful merchants who, when the heyday of shipping began to fade, lost no time transferring their capital and their wits to manufacturing."[9] Indeed, the growth of the city's commercial river and railroad trade created new wealth and capital reserves among Hartford's wealthy merchant class and the banks and insurance companies that served them. These profits, deposits, and premiums needed to be reinvested, and direct investments in local manufacturing ventures were one of the few outlets for excess capital at this time because Wall Street's capital markets were still emerging in the mid-1800s.[10]

Beyond its investment capital, Connecticut also boasted some extraordinary human capital as a hotbed of inventive talent. From the initiation of the U.S. patent system in 1790 through about 1930, Connecticut led all states in the number of patents granted per capita. Beyond Hartford, a short list of notable Connecticut inventors includes Eli Whitney (Hamden, New Haven), an arms maker and inventor of the cotton gin; clock maker Eli Terry (Windsor, Plymouth); Charles Goodyear (New Haven, Naugatuck), inventor of vulcanized rubber; Elias Howe (Bridgeport), one of the inventors of the lockstitch and modern sewing machine; submarine inventor David Bushnell (Saybrook); and Linus Yale Sr. (Middletown), inventor of the pin-tumbler lock. When historians speak of Yankee ingenuity, this is what they mean. Thus, Hartford's inventors were embedded in a generally stimulating environment that encouraged tinkering, risk-taking, and entrepreneurship.[11]

PEOPLE
Hartford's Mechanics

Hartford native Samuel Colt (1814–62) is unques-
tionably the key figure in the city's ascendance as
a place of invention and manufacturing powerhouse (Figure 2). Colt is perhaps best known for
inventing his namesake revolver, but he was also
a daring entrepreneur, a dynamic salesman on a
global scale, and one of America's wealthiest men.
However, Colt's greatest innovation was developing a precision manufacturing system to produce
his weapons on a mass scale. In the process, Colt's
armory and the surrounding Coltsville community became a training ground for dozens of
skilled machinists, who would apply these techniques in several other Hartford industries.

In 1830, the teenaged Colt left Amherst Academy for a job as a common seaman on the *Corvo*,
a brig headed for Calcutta. According to legend,
Colt watched the ship's wheel turn and lock in
position with a clutch; this inspired him to whit-
tle a wooden model of a so-called pepperbox pis-
tol with a single chamber and multiple revolving
barrels, an already common design. As his think-

FIGURE 2
A portrait of inventor Samuel Colt, 1859. PG 460,
Colt's Patent Fire Arms Manufacturing Company
collection, State Archives, Connecticut State Library.

ing progressed, Colt inverted the pepperbox design; his new concept featured a single barrel and
a revolving cylinder with multiple chambers that rotated into place when the shooter cocked the
hammer.[12]

After obtaining U.S. Patent 9430X in 1836, Colt secured backing from family members and
some New Jersey capitalists and opened his first manufacturing facility in Paterson, New Jersey.
Colt's revolvers were used in the fighting for Texas independence and Florida's Seminole Wars.
However, the Paterson factory produced revolvers of inconsistent quality, and with these production problems, slack sales, and tight credit after the 1837 financial panic, Colt's Paterson venture
went belly-up around 1842. After a hiatus, Colt was contacted in 1845 by Captain Samuel Walker,
whose Texas Rangers wanted multishot pistols they could fire from horseback during the Mexican
War. Walker ordered 1,000 revolvers, but Colt was out of business; undaunted, Colt contracted
with Eli Whitney Jr. of Hamden, Connecticut, in 1847 to manufacture Walker's order. Whitney

was a wise choice. His father, of course, was Eli Whitney Sr., who is best known as the inventor of the cotton gin but was also one of the first private arms makers to attempt interchangeable parts manufacturing outside the federal armories. The younger Whitney's superintendent was Thomas Warner, the former chief arms maker at the federal Springfield Armory, just 30 miles up the Connecticut River from Hartford.

Whitney and Warner fulfilled that initial order and inspired Colt to improve his own manufacturing processes. With more orders in the pipeline, Colt moved back to Hartford and rented temporary factory space, first on Pearl Street (1847–49) and later on Commerce Street (1849–55).[13] Colt needed a superintendent, and he hired Elisha K. Root (1808–65) away from the Collins and Company axe works of Canton, Connecticut, where he had mechanized the production process.[14] Colt and Root made a formidable team. Colt was the face of the company, a savvy marketer and charismatic salesman who schmoozed U.S. Army generals and built a global business, selling firearms to armies in Britain, Russia, the Ottoman Empire, and India. Meanwhile, Root handled the technical operations and delivered the product back at the armory. According to historian Joseph Wickham Roe, "the credit for the revolver belongs to Colt; for the way they were made, mainly to Root."[15]

With sales booming, Colt began developing plans for a new, permanent factory in Hartford. The resulting Colt Armory and surrounding Coltsville industrial village were key places of invention within Hartford and a testament to Samuel Colt's audacious vision and perseverance. In 1853, Colt purchased 200 acres of floodplain at South Meadow along the banks of the Connecticut River and self-financed the construction out of his own enormous profits. To protect against floods, Colt employed 100 workers to build a dike 2 miles long, 10 to 20 feet high, and 40 feet wide across the top and then planted deep-rooted French osier (willow) trees to secure the earthen works. The main armory was three stories high and was laid out like an H; a star-spangled blue onion dome adorned with a bucking, rampant colt graced the rooftop. The main factory floor was 500 feet long, 60 feet wide, and 16 feet high, with 60 cast-iron columns supporting the steam-powered drive shafts and pulleys that ran the machinery (Figure 3).[16]

According to the 1860 U.S. census, the armory employed 369 workmen and turned out 45,000 firearms a year. With a capitalization of $1.25 million, the Colt Armory was surely one of the largest and most sophisticated manufacturing operations in the late nineteenth century.[17] The armory was badly damaged (and rebuilt) after a suspicious fire set it ablaze in 1864 at the height of Civil War production.[18] In 1868, Mark Twain toured the reconstructed factory and observed,

> On every floor is a dense wilderness of strange iron machines . . . a tangled forest of rods, bars, pulleys, wheels, and all the imaginable and unimaginable forms of mechanism. There are machines to cut all the various parts of a pistol, roughly, from the original steel: machines to trim them down and polish them; machines to brand and number them; machines to bore the barrels out; machines to rifle them;

FIGURE 3
Colt's east armory workers with their Lincoln milling
machines, about 1900. Notice the metal shavings at their feet.
PG 460, Colt's Patent Fire Arms Manufacturing Company
collection, State Archives, Connecticut State Library.

machines that shave them down neatly to a proper size, as deftly as one would
shave a candle in a lathe.

As Twain concluded, "One can stumble over a bar of iron as he goes in at one end of the establishment, and find it transformed into a burnished, symmetrical, deadly 'navy' [model revolver] as he passes out at the other" (Figure 4).[19]

However, the armory was merely the centerpiece of the Coltsville industrial complex, which included a detached three-story office with an ornamental clock tower, a blacksmith's shop, a proving house for testing firearms, and a dozen other scattered buildings (Figure 5). Always the entrepreneur, Colt recruited artisans from Europe, built a series of German-style chalets in a section deemed Potsdam Village, and set the craftsmen to work making furniture from the willow branches that fell along his riverside dikes. Beyond these workplaces, Coltsville featured Armsmear, Colt's ornate Italianate mansion, and housing for some 145 families. Colt also constructed baseball fields and Charter Oak Hall, a 100-foot-long, four-story brick building "devoted to purposes of moral, intellectual, and artistic culture." Charter Oak Hall featured classrooms where immigrant workers learned English; a library and lecture hall; a banquet space for socials, dances, and other

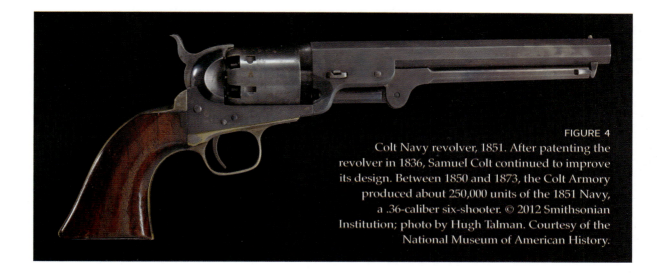

FIGURE 4

Colt Navy revolver, 1851. After patenting the revolver in 1836, Samuel Colt continued to improve its design. Between 1850 and 1873, the Colt Armory produced about 250,000 units of the 1851 Navy, a .36-caliber six-shooter. © 2012 Smithsonian Institution; photo by Hugh Talman. Courtesy of the National Museum of American History.

civic events; and a series of activity rooms that hosted the Colt Bicycle Club and rehearsals of the Colt's Armory Band.[20]

Colt suffered from inflammatory rheumatism and, at age 47, died in 1862 at the peak of Civil War production. His trusted superintendent Elisha K. Root took over as president (Figure 6). Root was widely regarded as the finest mechanic in New England, and the success of the armory attracted skilled machinists from all over the northeast. Root hired dozens of "inside contractors," contract arms makers who used Colt's shop space, power, machine tools, and materials to produce a particular part or to manage a particular operation for a set piece rate. Inside contracting encouraged innovation because competing contractors strived to improve their techniques and lower their costs to win Colt's business. As Root shared his expertise, Colt's armory became known as a "college of mechanics," where skilled machinists and contractors built and operated Colt's specialized machine tools and mastered the various aspects of the manufacturing process.[21]

As they gained in knowledge, these ambitious mechanics and contractors often left Colt's armory to apply the techniques of "armory practice" with other clients or to start their own new firms. This spin-off phenomenon—in which ambitious inventors and entrepreneurs train first at a leading firm before leaving to found their own start-ups—is characteristic of many places of invention. For example, both Fairchild Semiconductor in Silicon Valley, California, and Medtronic in Medical Alley, Minnesota, spawned several new start-ups and a shared loyalty among former employees.

Similarly, Colt's former employees and contractors formed a tight-knit community, marked by several intertwined relationships and high job mobility that facilitated the cross-pollination of technical knowledge across Hartford's various industries.[22] For example, Charles E. Billings (1835–1920) worked for Colt as a contract die maker and die forger from 1854 to 1856. While at Colt's

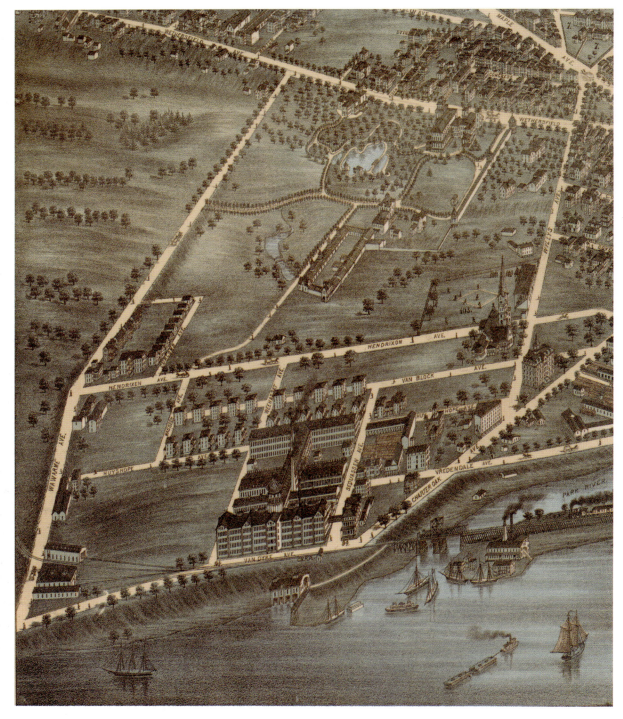

FIGURE 5

Detail from *The City of Hartford, 1877* by lithographer O. H. Bailey, showing a bird's-eye view of Coltsville. This industrial village along the Connecticut River was protected by willow-lined dikes and included the famous onion-domed factory (foreground) and, behind it, workers' housing, a baseball field, and a church. Sam and Elizabeth Colt's Italianate mansion, Armsmear, is at the top right. The Connecticut Historical Society.

armory, he met fellow contractors George Fairfield (?–1908) and Christopher M. Spencer (1833–1922). Spencer was the ultimate mobile mechanic, a classic Yankee tinkerer and entrepreneur who worked across several sectors. After leaving Colt's armory in 1856, Spencer went to work at the Cheney Brothers silk mills in Manchester, across the Connecticut River from Hartford. At Cheney Brothers, he invented an automatic winding machine that spun silk onto a spool; the machines were built by two former Colt employees, Francis Pratt (1827–1902) and Amos Whitney (1832–1920), whose Hartford-based Pratt & Whitney became the most important machine tool firm in the northeast and one of Hartford's biggest employers (Figure 7).[23]

While moonlighting from Cheney Brothers, Spencer patented a breech-loading repeating rifle and formed the Spencer Repeating Rifle Company in 1861

FIGURE 7
The 35 employees of Pratt & Whitney, 1886. Francis A. Pratt and Amos Whitney, seated front row center, met as inside contractors at Colt's armory and started their own spin-off firm in 1860. Source: Pratt &Whitney, *Accuracy for Seventy Years, 1860–1930* (Hartford, Conn.: Pratt &Whitney, 1930), 24. © 2014 Smithsonian Institution; photo by Richard Strauss. Courtesy of Smithsonian Institution Libraries.

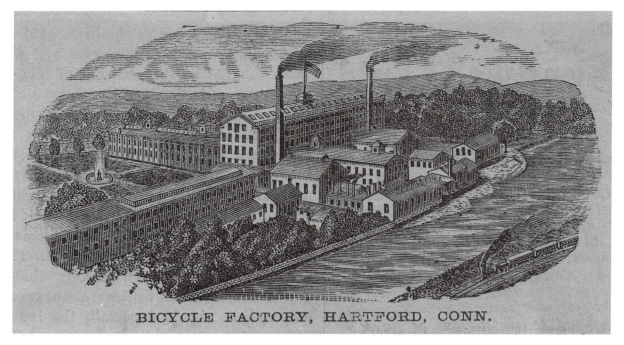

BICYCLE FACTORY, HARTFORD, CONN.

FIGURE 8

The factory at 436 Capitol Avenue, Hartford, 1881, which, at various times, produced several different products under the same roof. Source: "A Great American Manufacture," *Bicycling World and Archery Field* 2 (1 April 1881): 326. © 2014 Smithsonian Institution. Courtesy of Smithsonian Institution Libraries.

to manufacture it. Spencer's repeater was tested personally by Abraham Lincoln and was adopted widely during the Civil War. In 1869, Spencer reunited with Charles Billings in Hartford to form Billings & Spencer, a firm that specialized in drop forgings and hand tools such as wrenches and pliers. Spencer later invented an automatic screw-making machine and in 1876 teamed up with Fairfield to found the Hartford Machine Screw Company. Fairfield had previously worked with Billings at Hartford's Weed Sewing Machine Company, where they made sewing machines, bicycles, and threaded screws all under the same roof at 436 Capitol Avenue (Figure 8).[24]

The multiple occupants of this single factory and the wide variety of products manufactured there underscore the general applicability and wide adoption of armory practice across a variety of Hartford industries. The story begins with yet another Hartford's arms maker, Christian Sharps (1810–74). In the 1830s, Sharps had apprenticed at the Harper's Ferry armory under John Hall, inventor of the first breech-loading rifle. While at Harper's Ferry, Sharps observed some of the deficiencies in Hall's breech loader and

The works of the Weed Sewing Machine Company cover two acres of ground, and besides manufacturing the sewing machine, they make bicycles and an extensive line of fine steel and iron forgings.

"THE MANUFACTURE OF SEWING MACHINES," SCIENTIFIC AMERICAN

FIGURE 9
Inventor Christian Sharps submitted this prototype breech-loading rifle to secure U.S. Patent 5763 in 1848. © 2014 Smithsonian Institution; photo by Jaclyn Nash. Courtesy of the National Museum of American History.

learned many of the techniques of interchangeable parts manufacturing. Sharps earned a patent for his own improved breech-loading rifle design in 1848 and contracted with toolmakers Samuel Robbins and Richard S. Lawrence of Windsor, Vermont, to manufacture it. In 1852, Robbins & Lawrence opened a new satellite factory at 436 Capitol Avenue in Hartford to build their own tools and fulfill Sharps's contract for 10,000 breech loaders. Sharps's rifles and carbines were extremely popular because of their accuracy; some 100,000 were manufactured during the Civil War and helped popularize the term *Sharps Shooters* (Figure 9).[25]

After the Civil War, Sharps also manufactured Weed sewing machines alongside his rifles. Historians know very little about inventor Theodore E. Weed, who died before his sewing machine patent was issued in 1854. Apparently, some investors acquired the rights to manufacture Weed's patented sewing machines, first in Nashua, New Hampshire, then in Saint John, New Brunswick, before moving their operations to Hartford. From 1865 to 1871, the Weed Sewing Machine Company contracted with Pratt & Whitney to manufacture their Family Favorite machines (Figure 10) and other models. Eventually, Weed needed more capacity, so in 1871 the firm leased space in the larger Sharps factory. When Sharps changed ownership in 1875 and moved to Bridgeport, Weed purchased the factory outright and took over operations

FIGURE 10
Weed Family Favorite sewing machine, about 1870s. © 2014 Smithsonian Institution; photo by Richard Strauss. Courtesy of the Museum of Connecticut History.

at 436 Capitol Avenue. Weed likely retained many of Sharps's employees, who could just as easily build guns or sewing machines.[26]

As noted earlier, George A. Fairfield, a former Colt contractor, served as superintendent (1865–76) and later president (1876–81) of the Weed Sewing Machine Company. After the Civil War, Fairfield lured his former Colt associate Charles E. Billings to help with Weed's operations, and together they built the firm's reputation for skilled machinists and efficient production. In a now familiar pattern, other inventors and entrepreneurs sought out Weed as a contract manufacturer. As noted earlier, Christopher Spencer approached Fairfield with his automated screw machine, and together they formed the Hartford Machine Screw Company in 1876, which initially operated out of a spare room at the Weed factory. Later, as the business grew, Fairfield left Weed in 1881 to become president of Hartford Machine Screw, which eventually built its own factory down the street at 476 Capitol Avenue.[27]

While still at Weed, Fairfield received another interesting pitch from a Boston shoe materials supplier named Albert A. Pope (1843–1909). During a visit to the 1876 Centennial Exposition in Philadelphia, Pope witnessed demonstrations of a high-wheel bicycle, a novelty from Britain. Sensing a commercial opportunity, Pope began importing bicycles from England and hatched a plan to produce them domestically. So it was in 1878 that Pope rode the train from Boston to Hartford, then, to the amazement of the city's onlookers, rode his high wheeler from the station down Capitol Avenue to the Weed Sewing Machine Company. Pope approached George Fairfield with a proposal: would Weed agree to build a test run of 50 bicycles under contract? When Fairfield agreed, Pope (via the Weed Sewing Machine Company) became the first domestic manufacturer of bicycles in the United States (Figure 11).[28]

The Pope Manufacturing Company, maker of Columbia brand bicycles, thrived under its owner's shrewd leadership. Pope bought up several controlling bicycle patents, which enabled him to prevent manufacturing by certain competitors while charging royalties and licensing fees to others. He positioned literally thousands of sales agents across the United States and overseas and developed artistically sophisticated catalogs and advertisements, which ran in national magazines. Pope was also instrumental in founding the League of American Wheelmen in 1880 and thereafter led the so-called Good Roads movement, which encouraged legislators to invest in the infrastructure necessary for bicycles (and, later, automobiles; Figure 12).[29]

Most importantly, Pope encouraged innovation in his product and production processes. The high-wheel "ordinaries" were tall, heavy, and notoriously difficult to ride, which confined their sales to vigorous and daring young men (Figure 13). Thus, in 1886, Pope and his engineers introduced the now familiar "safety" bicycle with two equally sized wheels, hollow steel frames, and pneumatic rubber tires, which expanded the bicycle market to women and children. Meanwhile, Pope engineers such as Hayden Eames and Henry Souther pioneered new manufacturing techniques such as metal stamping and electrical resistance welding while maintaining the industry's

FIGURE 11
Machining rear hubs for Pope's Columbia brand bicycles, inside the Weed Sewing Machine Factory, 1881. Source: "A Great American Manufacture," *Bicycling World and Archery Field* 2 (1 April 1881): 326. © 2014 Smithsonian Institution. Courtesy of Smithsonian Institution Libraries.

FIGURE 12
A glass paperweight featuring Albert A. Pope, leader of the American Good Roads movement, about 1890s. © 2014 Smithsonian Institution; photo by Richard Strauss. Courtesy of National Museum of American History.

COL. ALBERT A POPE
FOUNDER OF THE
MANUFACTURE OF BICYCLES
IN THE UNITED STATES
AND THE
PIONEER OF THE GREAT MOVEMENT
FOR BETTER AMERICAN ROADS.

FIGURE 13
Columbia Light Roadster, 1886. © 2014 Smithsonian Institution; photo by Hugh Talman. Courtesy of National Museum of American History.

leading testing and quality assurance department. Sales boomed, and Pope eventually gained control of Weed in 1881; sewing machine production continued alongside bicycles for another decade until Pope phased out the sewing machines in 1891 to focus exclusively on bicycles. By 1895, Pope's expanded Hartford operations included five factories set on 17 acres, employing 4,000 workers, making him Hartford's largest employer.[30]

Pope's product and process innovations drove the price of bicycles down to around $100–$125, and soon the "bicycle craze" saturated the market. Thus, Pope began casting about for new products and recruited a young, Massachusetts Institute of Technology–trained inventor named Hiram Percy Maxim (1869–1936) to experiment with motorized bicycles and carriages.[31] While working for an armaments company in Lynn, Massachusetts, Maxim took long bicycle rides to visit a girlfriend in a neighboring town, which led him to experiment with connecting a small gasoline motor to a tricycle. Maxim's friend Hayden Eames arranged a demonstration for Albert Pope in Hartford, and in 1895 Maxim was hired to head up Pope's new Motor Carriage Division. Automobiles would soon join machine tools, rifles, sewing machines, threaded screws, and bicycles on the list of items manufactured at 436 Capitol Avenue.[32]

Maxim developed both gasoline and electric vehicles for the Pope Manufacturing Company.[33] At first glance, Maxim's "horseless carriages" borrowed much from Pope's bicycles; for example, both vehicles utilized gears, pneumatic tires, and metal tubing for frames and axles, and both operated more smoothly on good roads. Nevertheless, as Maxim recalled in his memoir, "Everything had to be created. There were no suitable bearings, suitable wheels, tires, batteries, battery-handling equipment, battery-charging equipment, gasoline engines, carburetors, spark plugs, brakes, or steering gears."[34] Thus, Maxim's motorized carriages cemented several conventions we now take for granted:

> Maxim, I believe this horseless-carriage business will be one of the big businesses of the future.
>
> ALBERT A. POPE
> TO HIRAM PERCY MAXIM,
> CHIEF OF POPE'S MOTOR
> CARRIAGE DEPARTMENT

a front-mounted engine, a multigear power train, and the driver's steering wheel on the left side (Figure 14). As Maxim's designs stabilized, production scaled rapidly. In 1897 and 1898, Pope manufactured about 500 electric and 40 gasoline carriages, far outpacing other upstarts like the Duryea Brothers and Michigan's Ransom E. Olds. Thus, Hartford—not Detroit—held the early lead in becoming America's automotive capital (Figure 15).[35]

However, Hartford was not destined to be Motown. Pope had borrowed heavily to finance his expansion into automobiles, and after the financial panic of 1907, the Pope Manufacturing Company went into receivership. The colonel died in 1909, and when the courts and creditors sorted everything out, the factory at 436 Capitol Avenue was liquidated and sold to Pratt & Whitney.[36] Meanwhile, Pope's engineers dispersed and became leaders in the maturing automobile industry; for example, Hayden Eames left Hartford and became the general manager at Studebaker. Hiram Percy Maxim also continued inventing. Inspired by the mufflers he had developed to quiet his internal combustion engines, Maxim invented the gun silencer in 1909 and contracted with the Colt Armory for its production.[37]

To this point, we have focused on Hartford's industrial leaders and the highly skilled inventors and engineers who spearheaded the city's precision manufacturing industries. However, it is important to remember the city's unskilled laborers who tended the high-tech machine tools and built the revolvers and sewing machines that came down the assembly line. A major class division existed in Hartford; the industrialists and skilled engineers were typically native-born Protestant Yankees, complemented by an occasional German- or Swiss-trained machinist. In contrast, the city's laborers were typically Irish Catholic immigrants who had fled the midcentury famines in search of factory work in America. Conservative, Puritan Hartford was utterly transformed by Irish migration, as the city's population doubled to 30,000 residents in the decade between 1850 and 1860.[38]

FIGURE 15

This radiator emblem, 1903–14, adorned Pope's Pope-Hartford brand, a midpriced, gas-powered automobile manufactured in Hartford. © 2014 Smithsonian Institution; photo by Richard Strauss. Courtesy of National Museum of American History.

Factory work was often unpleasant, with long hours, relatively low pay, and the noise, fumes, and danger of working with heavy machinery. As noted, Samuel Colt invested in on-site workers' housing, baseball fields, the Colt's Armory Band, and several activities at Charter Oak Hall to retain workers and boost morale (Figure 16). Similarly, Pope built workers' housing and a recreational park adjacent to his Capitol Avenue factory; inside, his workers enjoyed 7¢ hot lunches and an employee library. The parks named for Colt and Pope are still present in twenty-first-century Hartford and are reminders of their industrial legacy, long after the factories have closed. However, it is important to remember that these workers' amenities were, at their root, paternalistic attempts by industrialists to pacify their labor forces and retain their loyalty through "welfare capitalism." It is also important to remember that these efforts had their limits because striking workers helped seal the fate of Pope Manufacturing in 1906 and temporarily crippled Colt in 1935.[39]

There is one final Hartford inventor who deserves our attention, and that is Mark Twain (Samuel L. Clemens, 1835–1910; Figure 17). After his 1868 visit, Twain moved his family to Hartford in 1871 to be closer to his publisher, Elijah Bliss Jr. of the American Publishing Company. Twain took residence in Hartford's famed Nook Farm literary community, where his neighbors included abolitionist Harriet Beecher Stowe, author of *Uncle Tom's Cabin,* and Charles Dudley Warner, editor of the *Hartford Courant* and Twain's coauthor on *The Gilded Age* (1873).[40] Twain was obviously inspired by Hartford's inventive community; for example, he based the mechanically gifted hero of

FIGURE 16
A trombonist in his Colt's Armory Band uniform, about 1850s. Museum of Connecticut History, Accession #tg1201.

FIGURE 17
Mark Twain, 1871. Detail of photo by Matthew Brady, Brady-Handy Collection, Prints and Photographs Division, Library of Congress, # LC-BH832-1426.

A Connecticut Yankee in King Arthur's Court (1889) on former Colt superintendent Elisha K. Root. In real life, Twain earned three patents in Hartford for an adjustable garment strap (1871), a prepasted scrapbook (1873), and a board game (1885). However, Twain was an unsuccessful venture capitalist and nearly went bankrupt after investments in several failed inventions, including James Paige's overly complex typesetting machine. In fact, Twain was forced to move his family from high-rent Hartford in 1891 to live less expensively in Europe and earn money on an extended speaking tour. With typical self-deprecating humor, Twain later suggested that "to succeed in business, avoid my example."[41]

INVENTION

The Application of Armory Practice

All kinds of things were invented and manufactured in Hartford, including firearms, sewing machines, bicycles, and automobiles. In addition, Hartford was a hub for the invention of machine tools—the drop hammers, milling machines, lathes, and drill presses—used to build other machines. Thus, although Colt, Pope, and their Hartford colleagues built many great products, their finest contribution was the development of the *manufacturing processes* to produce those products on a mass scale.

Hartford's ascendance as a leading manufacturing center began in the firearms industry, with the establishment of the Colt armory and the factory at 436 Capitol Avenue that first manufactured Christian Sharps's breech-loading rifle. The U.S. military, state militias, and local police departments were the primary customers in the firearms business, so government patronage basically subsidized the development of precision manufacturing expertise in Hartford. In addition, private arms makers like Sharps and Colt readily adopted innovations in both gun design and interchangeable parts manufacturing that had emerged from the federal armories in Harper's Ferry, Virginia (now West Virginia), and Springfield, Massachusetts, just 30 miles up the Connecticut River from Hartford. Thus, government (especially military) funding and know-how were crucial to the development of Hartford and many other places of invention. For example, Silicon Valley's early electronics industry grew steadily on the strength of military contracts for klystrons, transistors, and integrated circuits, whereas grants from the National Science Foundation have sustained Fort Collins researchers Bryan Willson and Amy Prieto at Colorado State University.[42]

Beginning in the early 1800s, federal arms contracts stipulated that all firearms be manufactured with uniform, interchangeable parts. Why was interchangeability so important? In traditional artisanal practice, a single highly skilled gunsmith would make and assemble every part of a musket, "lock, stock, and barrel." Each of the metal assemblies—the lock, the trigger, and the barrel—had to be hand filed so that they fit together properly. Thus, each musket was unique and made to order, so production volume was low and the unit cost of each gun was high. This method of artisanal production had certain military disadvantages. First, the slow pace of production made it difficult for the military to maintain a steady supply of arms. Also, if an artisan-produced musket or revolver failed on the battlefield, the Army would need a field blacksmith to repair it since each gun was unique. Theoretically, if identical copies of a master gun design could be produced with identical, interchangeable parts, a broken lock or trigger could be repaired quickly by simply changing out the part. To achieve uniformity, the federal armories at Springfield and Harper's Ferry began experimenting with specialized machine tools such as lathes and milling machine to forge the individual gun components using the division of labor.[43] The Army also wrote interchangeability into the requirements for all new gun

> The Town of Hartford, Connecticut is the greatest center of activity in the automobile industry today.
> BRITISH OBSERVER, 1897

orders when they contracted with suppliers like Colt and Sharps. As inspectors from the federal armories traveled among their contractors, they gathered and shared knowledge and best practices throughout New England.[44]

After struggling with production problems at his Paterson, New Jersey, factory, Samuel Colt achieved a breakthrough after hiring Elisha K. Root as the superintendent of his new Hartford armory. After joining Colt in 1849, Root systematically applied his deep knowledge of die forging, milling, grinding, and boring to the production of Colt's revolvers. Root and his workers developed precise molds for forging the basic metal pieces and weighted drop hammers to stamp them out. Root also invented all kinds of specialized lathes, drill presses, and milling machines to grind those metal blanks into the finished components (Figure 18). To achieve precision, Root also employed

FIGURE 18
Lincoln Milling Machine, 1861. Tool builder George S. Lincoln of Hartford's Phoenix Iron Works manufactured this type of general-purpose milling machine, according to the design of Colt's superintendent Elisha Root and machinist Francis A. Pratt. It was used in nearly every Hartford factory. Its high-speed rotary cutting blade shaved excess material from a metal work piece. © 2014 Smithsonian Institution; photo by Jaclyn Nash. Courtesy of National Museum of American History.

FIGURE 19

To ensure the interchangeability of individual gun components, inspectors at the Colt Armory tested each part with gauges like these, developed for the 1898 service revolver. © 2014 Smithsonian Institution; photo by Richard Strauss. Courtesy of the Museum of Connecticut History.

jigs and bearing points to secure the blank pieces on the cutting machines and a series of inspection gauges and calipers to ensure that the finished pieces conformed to exact specifications (Figure 19). These specialized machine tools—machines to make other machines—eliminated the variability introduced by hand forging and filing. Under Root's leadership, the techniques of armory practice first employed at Springfield and Harper's Ferry truly came of age. By employing division of labor,

Component Parts of Double Action Revolving Pistols.

.38 and .41 Cal.

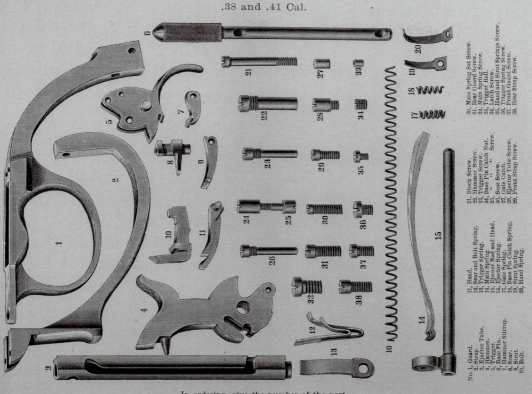

In ordering, give the number of the part.

(13)

FIGURE 20

"Component Parts of Double Action Revolving Pistols," from Colt's 1890 catalog. This exploded view and partial list of parts suggests the precision of Colt's manufacturing process. Source: Colt's Patent Fire-Arms Manufacturing Company, *Military and Sporting Arms* (Hartford, Conn.: Colt, 1890). © 2014 Smithsonian Institution; photo by Richard Strauss. Courtesy of Smithsonian Institution Libraries.

specialized machine tools, and precise quality control standards, Root and his team of skilled machinists were among the first private manufacturers to build guns with interchangeable parts and achieve volume production on a mass scale (Figure 20).[45]

As they moved on to lead firms in other industries, Colt-trained machinists demonstrated that the use of specialized machine tools, jigs, and inspection gauges had general applicability in any industry that required the precision forging, stamping, and milling of metal parts. In other words, once you understood the general techniques, forging and milling a revolver's trigger was not so different from forging and milling a sewing machine's bobbin or a bicycle's sprocket or an automobile's gear or a typewriter's key.[46] As Samuel Colt explained more simply in 1854, "There is nothing that cannot be produced by machinery."[47]

WHAT HAPPENED TO THE HOT SPOT?
HARTFORD IN THE TWENTY-FIRST CENTURY

Unfortunately, since its heyday in the late 1800s, Hartford has experienced a serious economic decline. In 2002, the *New York Times* described the current condition of Hartford's best-known industrial landmark (Figure 21):

> On the eastern edge of town, looming over the broad brown waters of the Connecticut River, the old Colt firearms factory is losing its fight with history. Under its spangled blue onion dome, the 19th-century armory where Samuel Colt's inventions helped spawn the Industrial Revolution—and where his company made the guns that helped the United States conquer the West and win two world wars— now stands hollow, disintegrating into crumbling bricks and broken glass.[48]

FIGURE 21
The Colt Armory and its trademark blue onion dome, 2011. After years of extensive lobbying by Hartford's city leaders, Congress and the U.S. Department of the Interior approved plans in December 2014 to turn the decaying Colt Armory and surrounding Coltsville neighborhood into a national park. © 2011 Smithsonian Institution; photo by Monica Smith.

After emerging as one of the world's great manufacturing centers of the nineteenth century, Hartford struggled during the twentieth century with deindustrialization, urban blight, white flight, crime, poverty, and underperforming schools. It is tempting to characterize Hartford's downturn as another familiar story of postindustrial decline, but the details are more complicated. As the 2000 census indicated, 30% of Hartford's residents lived below the poverty line, the second worst rate nationally. However, although inner-city Hartford continues to struggle, that same 2000 census listed the greater Hartford metropolitan area as having the nation's sixth highest median income.[49] What happened?

> It is difficult for young people today to imagine life without motorcars and good roads. Both are accepted as matters of course. Scant thought is given to the fact that they had to be created.
>
> HORSELESS CARRIAGE INVENTOR HIRAM PERCY MAXIM

In general, Hartford prospered through the 1940s and early 1950s. Although Pope had gone bankrupt in 1907, some new firms continued to move in. For example, two major typewriter companies, Underwood (1901) and Royal (1907), relocated to Hartford and made the city the typewriter capital of the world for half a century. Later, in 1925, Frederick Rentschler came to Hartford from the Wright Aeronautical Corporation and partnered with a key local firm to establish Pratt & Whitney Aviation (distinct from Pratt & Whitney Machine Tool) in the production of radial, air-cooled airplane engines. The region also enjoyed rising white and "pink" collar incomes from the growing insurance and financial industries, as well as the continuing growth in civil service jobs after Hartford was named the state capital in 1874. In general, during the early decades of the twentieth century, Hartford's population and economy were still among the most diverse, technologically sophisticated, and profitable in New England.[50]

The manufacturing exodus from inner-city Hartford started the city on its road to decline, and it can be traced to a peculiarity of the region's municipal politics. In the 1800s, the area known as Hartford covered about 87 square miles, spanning both sides of the Connecticut River and encompassing modern-day Hartford, West Hartford, East Hartford, and Manchester. However, beginning with West Hartford's split in 1854, the aforementioned towns broke away from Hartford, whose municipal borders eventually became fixed at only 18 square miles.[51] Eventually, the city's manufacturing firms needed more (and more modern) industrial workspaces than downtown Hartford's aging factories could easily provide. The suburban migration began almost immediately after Rentschler established his aerospace business in 1925, when Pratt & Whitney Aviation moved to a 1,000-acre site across the river in East Hartford, where it could build its own airport. Pratt and Whitney Machine Tool moved to West Hartford in 1939, followed by Colt in the early 1940s. Between 1951 and 1971, the number of factories within Hartford's city limits shrank from 41 to 16.[52]

Meanwhile, the suburbanization (or abandonment) of the factories also hastened the residential flight from inner-city Hartford. Hartford's population peaked in 1950 (177,397) but steadily declined over the next decade as an estimated 90,000 jobs left the city. Some, but not all, workers followed. Inspired by New Deal and post–World War II programs like the GI Bill that encouraged home ownership, workers and managers left their nineteenth-century downtown tenements for new suburban homes with modern plumbing and electricity. Unfortunately, redlining and other discriminatory loan policies fostered only white flight, leaving Hartford's growing African American and immigrant population to occupy a rapidly decaying city. Meanwhile, the growth of postwar car culture and the erection of Interstates 91 and 84 through Hartford encouraged further suburban sprawl and transformed Hartford into a city where people worked but did not live.[53]

In short, Hartford became what urban theorists describe as a city with "inelastic" boundaries, that is, one that cannot grow through annexation and thus becomes vulnerable to a vicious cycle. As an inelastic city loses its businesses and middle-class population to the suburbs, poverty becomes more concentrated in the inner city. With the high concentration of poor residents, the local government becomes squeezed, obligated to provide welfare, housing, and educational services, even as the local tax base has eroded. The city is often forced to raise taxes, driving even more local businesses and residents to the suburbs. City services decline—the schools, trash collection, and policing suffer—and the city becomes more crime ridden and dangerous. Thus, by the 1990s, Hartford had become what Tom Condon of the *Hartford Courant* called the "hole in the doughnut, the region's poorhouse, the tiny, impoverished core of a larger and wealthier region."[54]

> So, in the late 20th century, Hartford presented a double-face to the world: as an increasingly poor core city and a fundamentally prosperous metropolitan region.
>
> HARTFORD HISTORIAN
> ANDREW WALSH

Although pessimists can point toward corporate apathy, numerous poor planning decisions, and bad governance in Hartford's decline, optimists point out that the city's residual strengths—the insurance industry and the Connecticut state government—served as twin anchors of economic stability in the wake of deindustrialization. Fortunately, Hartford is making steps toward a recovery. Since the early 2000s, the state of Connecticut has injected some $400 million into several revitalization projects: building a new Connecticut Convention Center; erecting a new museum, the Science Center of Connecticut; renovating the Hartford Civic Center (now XL Center); and converting the old G. Fox department store building into Capital Community College. Finally, after years of extensive lobbying by Hartford's city leaders, Congress and the U.S. Department of the Interior approved plans in December 2014 to turn the decaying Colt Armory and surrounding Coltsville neighborhood into a national park.[55]

Despite the city's difficult economic circumstances, residents are hopeful that a renewed spirit of innovation can again flourish in Hartford, much as the vibrant creativity of hip-hop emerged from Bronx's impoverished "war zone" in the late 1970s. Creativity, risk-taking, and entrepreneurship run deep in Hartford's DNA, and the city's leaders have drawn inspiration from the city's innovative past as they try to muster a comeback.[56] However, Hartford's spectacular ascent, long decline, and attempts at revitalization remind us that places of invention rarely last forever and that each innovative region has a life cycle driven by its own particular circumstances.

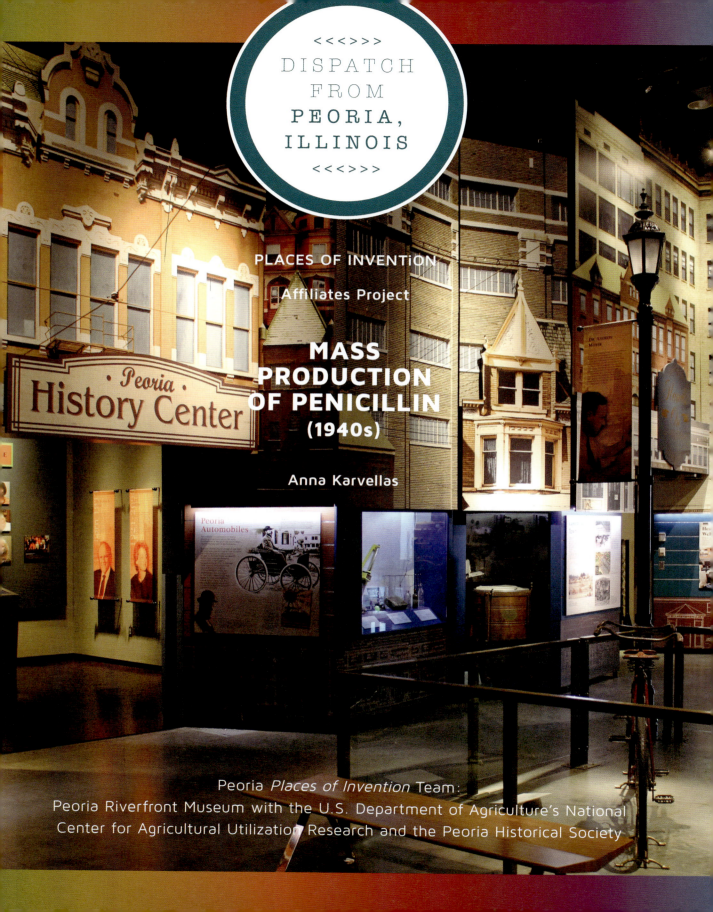

DISPATCH FROM PEORIA, ILLINOIS

‹‹‹›››

‹‹‹›››

PLACES OF INVENTION

Affiliates Project

MASS PRODUCTION OF PENICILLIN
(1940s)

Anna Karvellas

Peoria *Places of Invention* Team:
Peoria Riverfront Museum with the U.S. Department of Agriculture's National
Center for Agricultural Utilization Research and the Peoria Historical Society

A Peoria museum, government lab, and historical society came together to explore their city as a place of invention and to bring community-wide attention to a shared history of scientific innovation. With access to robust archives, the team of committed museum specialists, scientists, historians, and educators tell the story of Peoria's inventive breakthroughs in the mass production of penicillin and the reasons why Peoria was perfectly poised to play this part. It is a place-based story with global implications.

THE MUSEUM AND THE RIVER

When the Peoria Riverfront Museum (PRM) opened in 2012, one of its priorities was to connect the museum, its collections, and the community more directly to the Illinois River and the innovation it continues to drive (Figure 1). The museum was born out of the former Lakeview Museum of

OPENING IMAGE (OPPOSITE PAGE)
The Street exhibition, Peoria Riverfront Museum, Peoria, Illinois, 2015. Photograph by Jim Dwyer. Courtesy of the Peoria Riverfront Museum.

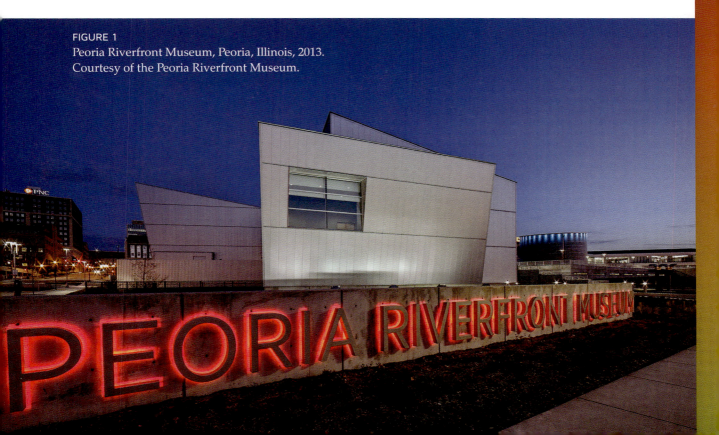

FIGURE 1
Peoria Riverfront Museum, Peoria, Illinois, 2013.
Courtesy of the Peoria Riverfront Museum.

Arts and Science and a public-private partnership of twelve organizations. Together, these groups worked over a decade to create a state-of-the-art museum along the Illinois River.[1] Its expansive environmentally-sustainable space was built next to a new Caterpillar, Inc., visitors' center in what the town hopes will become a revitalized waterfront cultural district anchored by these two local institutions. The nearby brick warehouses transformed into lofts and galleries speak to this potential for change and are a reminder of the days when Peoria was the bustling whiskey capital of the world. For most of Peoria's history, this spot was a center of industry and civic life—a place where people gathered and water, road, and rail met.

The museum's institutional realignment with the Illinois River makes sense: the river is intertwined with Peoria's evolution as a city and place of invention. One of the PRM's anchoring exhibitions, *Illinois River Encounter*, takes visitors through 14,000 years of the river's natural history, beginning with the Kankakee Torrent, a catastrophic flood that shaped the Illinois River Valley. Alongside this story, the exhibition examines the ways that people and the river have shaped each other over time in this central Illinois region (Figure 2).

This point was especially salient during the Smithsonian's visit in the aftermath of Peoria's record-breaking flood in 2013. The Illinois River crested 29.35 feet, leaving dark, rich tendrils of river mud as it receded from points just below the museum's sandbags.[2] The 2013 water level surpassed even that of the 1943 flood documented in the exhibition's flood model and oral histories, a reminder that the Illinois River continues to write—and rewrite—Peoria history.

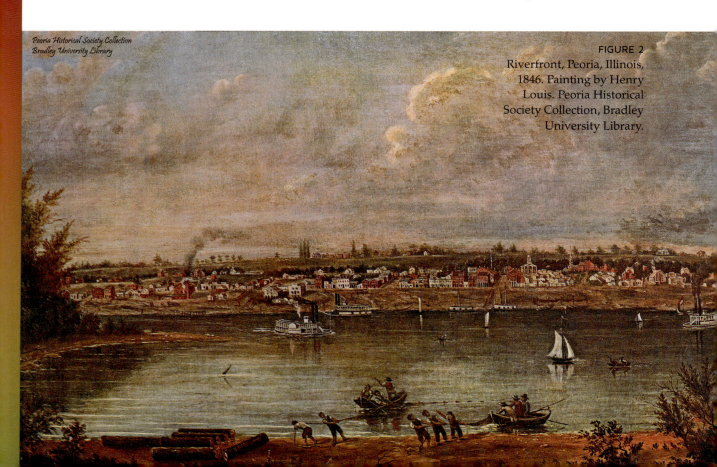

FIGURE 2

Riverfront, Peoria, Illinois, 1846. Painting by Henry Louis. Peoria Historical Society Collection, Bradley University Library.

SELECTING A PLACE-BASED INVENTION TOPIC

As the PRM and its partners began to develop content for the Affiliates Project, it was clear that the city's river, agricultural resources, and fermentation expertise would play a part. After attending the Smithsonian training workshop in Washington, they decided upon the city's most famous invention story: Peoria's development of processes for mass producing penicillin at what is now the U.S. Department of Agriculture's National Center for Agricultural Utilization Research (USDA-NCAUR), one of the partners on the project.

It is a captivating story about contingencies and consequences: how Peoria's Northern Regional Research Laboratory was ideally positioned to take British bacteriologist Alexander Fleming's discovery of penicillin, further advances made at Oxford University, and invent a process for mass producing penicillin that could be scaled up by pharmaceutical companies during World War II. Time was of the essence as Allied forces waited for an adequate penicillin supply to treat wounded soldiers. While the war raged, the Peoria lab, like other labs around the country, took samples of the British strain and conducted experiments to increase yield. It was in Peoria that fermentation expertise, experimentation, diligence, and luck combined to bring forth significant breakthroughs that would lay the foundation for commercial production of penicillin. The Peoria team introduced improved culture mediums and processes for deep fermentation and found a more productive strain of the *Penicillium* mold. In short, they discovered the best media for penicillin production, the best strain to use, and the best conditions to grow it.[3] As part of the international penicillin effort, Peoria scientists helped save countless lives that would have otherwise been lost as a result of infections (Figure 3). The discovery and development of penicillin is one of the twentieth century's most important scientific breakthroughs.[4]

> Penicillin, perhaps the outstanding life saving drug of all time, was a joint Anglo-American development. It was an example of cooperation between British and American scientists early in the war and before a basis for such community of effort was officially formalized. Our enemies developed no similar antibiotic and consequently suffered a higher mortality rate among their wounded.[5]
>
> —Dr. Robert D. Coghill, Northern Regional Research Lab

The selection of this topic dovetailed well with the PRM's work on its exhibition *The Street*, which included the penicillin story as part of its look at 300 years of invention and innovation in Peoria (opening image). *Places of Invention* (*POI*) training and funding allowed the museum to extend and add depth to their exploration of the topic. It also provided another opportunity for the museum to strengthen its relationship with the USDA-NCAUR, its primary community partner. Museum and programming staff were particularly eager to work with lab researchers to develop

FIGURE 3

"Thanks to Penicillin . . . He Will Come Home!" advertisement for Schenley Laboratories, 1944. Research and Development Division, Schenley Laboratories, Inc.

related science-based activities that could be presented independently or as part of *POI*-themed programs. These programs would connect historic and contemporary scientific invention in Peoria and would inspire creativity and innovation in participants.

> We have wanted to work with the NCAUR for so long, and *Places of Invention* has provided the perfect chance to do that.[6]
>
> —Kate Neumiller Schureman, Peoria Riverfront Museum

The National Center for Agricultural Utilization Research, in turn, was eager to share this local history to reconnect with a community that had in many ways been cut off from the lab when post-9/11 security measures ended public tours. Indeed, many Peorians no longer understood the kind of work the lab was doing in its historic complex on land once leased from Bradley University. By the time scientific breakthroughs went to market, their origins in Peoria government labs were not apparent to the average person.

A third partner was recruited to join the *POI* team, the Peoria Historical Society, a longtime collaborator and key partner in the development of the PRM.[7] The Peoria Historical Society's digital expertise and archival collections helped the team research and create video deliverables for the *POI* exhibition's interactive map. It had already produced an online exhibition about the penicillin story working with the USDA-NCAUR.[8]

As they moved ahead, the *POI* team met regularly and used "mind-mapping" software to track relationships between the people, places, and resources in their local invention and innovation stories. The tool helped them look closer at the Peoria penicillin story and recognize the influence of similar factors in other inventions. With careful planning, the group determined a way to create *POI* videos that could be incorporated into *The Street*'s look at Peoria innovation. The basic *POI* video structure would become a template for future innovation stories, including other stories outside of the USDA-NCAUR. Over time, the exhibition could potentially include other hometown innovators, such as the Avery and Caterpillar tractor companies and different branches of the medical industry.[9]

FOCUSING THE LENS
Why There? Why Then?

"This is a story about convergence," the *POI* team explained when the Smithsonian visited the USDA-NCAUR lab. Peoria's chapter of the penicillin story begins with the river and the city's rise as a center of agriculture and fermentation and distillation expertise. Peoria was a natural gathering spot for people to settle along the shores of the Illinois River. With fertile soil, plentiful grain, cheap coal, good water, and abundant wood, the region was especially attractive to European immigrants eager to turn their fermentation and distillation expertise into profit.

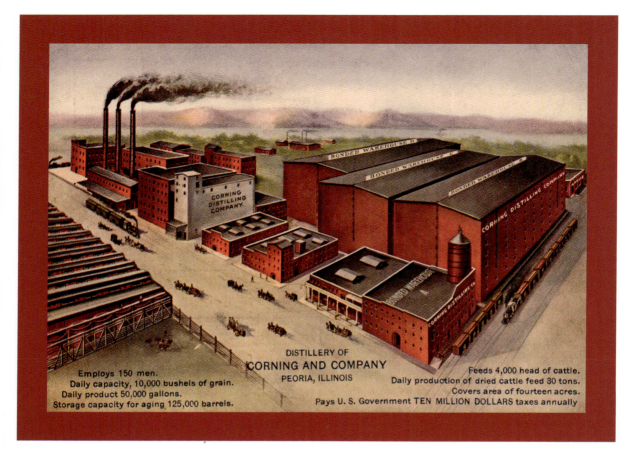

Employs 150 men.
Daily capacity, 10,000 bushels of grain.
Daily product 50,000 gallons.
Storage capacity for aging 125,000 barrels.

DISTILLERY OF
CORNING AND COMPANY
PEORIA, ILLINOIS

Feeds 4,000 head of cattle.
Daily production of dried cattle feed 30 tons.
Covers area of fourteen acres.
Pays U. S. Government TEN MILLION DOLLARS taxes annually

FIGURE 4
Distillery of Corning and Company, Peoria, Illinois, about 1880–1919. Peoria Historical Society Collection, Bradley University Library.

Access to a network of rails, rivers, and roads provided additional advantages and made Peoria a brewery and distillery boom town (Figure 4). An abundance of locally grown corn, barley, and rye was quickly transformed into whiskey and shipped to nearby Chicago and points beyond. Experimentation with fermentation and distillation techniques boosted quality and quantity. Between 1837 and Prohibition, the city was home to 24 breweries and 73 distilleries and drove a booming local economy that also was one of the country's largest grain markets. To keep up with demand, new tractors and other kinds of agricultural machinery were developed to increase productivity. Stockyards multiplied as cattle were fattened on leftover whiskey mash. A yellow haze and the mash's pungent smell hung over Peoria on humid summer evenings (Figures 5 and 6).[10]

Peoria, located between the industrial East and the agricultural West at the intersection of both natural and manmade transportation systems, was the ideal location for a "boom" economy. Finished iron products flowed west from the eastern

Places of Invention

FIGURE 5
Stockyards and distilling district, Peoria, Illinois, about 1914. United Postal Company. Peoria Historical Society Collection, Bradley University Library.

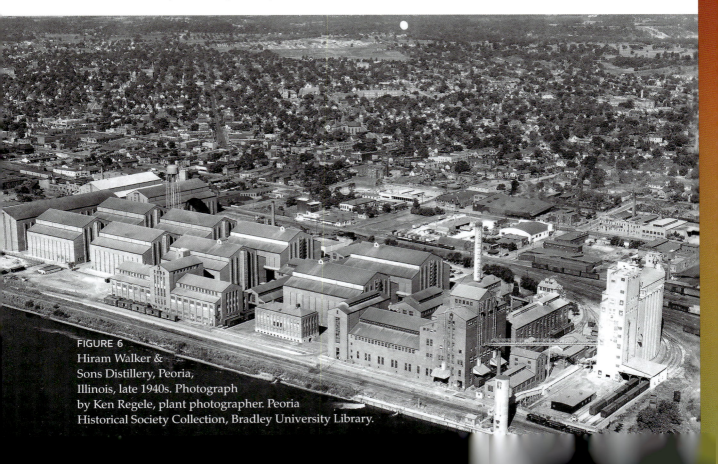

FIGURE 6
Hiram Walker &
Sons Distillery, Peoria,
Illinois, late 1940s. Photograph
by Ken Regele, plant photographer. Peoria
Historical Society Collection, Bradley University Library.

part of the United States while grain and lumber moved from the West to the East. These raw materials were utilized for the brewing and distilling industries that were seemingly tailor-made for Peoria.[11]

—Bryan J. Ogg, Peoria Historical Society

This prosperity did not last forever. Prohibition hit Peoria's alcohol industry hard, as did the Great Depression, which occurred as hybrid corn and advances in mechanization allowed for greater production.[12] By the late 1930s, the grain market could not keep up and was flooded with an abundance of product. It was in this context that Congress passed the 1938 Agriculture Adjustment Act, authorizing the construction of four regional research labs charged with "finding new uses and markets for farm commodities" with a focus on "corn, wheat and agricultural waste products" at the Peoria lab.[13] Congress appropriated $1 million to construct the lab on land leased from what is now Bradley University (Figure 7). The lab opened in December 1940. Its staff included fermentation experts who had worked together at the Bureau of Chemistry and Soil's Color Laboratory in Arlington, Virginia.[14] To aid the lab's work, partnerships with Peoria's local brewing and distilling industry provided a steady supply of corn-steep liquor, a production by-product that promoted microorganism growth.[15]

Seven months later, expertise and opportunity converged in Peoria when British scientists Dr. Howard Florey and Dr. Norman Heatley arrived seeking a lab to pursue the commercial production

FIGURE 7
Northern Regional Research Laboratory, Peoria, Illinois, about 1945. Collections of the USDA-NCAUR and Peoria Historical Society.

FIGURE 8

Penicillin team at the Northern Regional Research Laboratory, Peoria, Illinois, 1940s. Collections of the USDA-NCAUR and Peoria Historical Society.

of penicillin. Conditions in a besieged London necessitated a globalized effort for Allied countries desperate to find a better-producing strain of penicillin. A former Color Lab scientist, Dr. Percy Wells, had met Florey and Heatley in Washington and recognized the lab's appropriateness to their mission. Wells knew that the Peoria lab's Fermentation Division, under the direction of Dr. Robert Coghill, could provide the mold cultivation expertise and an engineering perspective that would help advance the British scientists' work (Figure 8).[16]

Among the lab's team of scientists and technicians were chemist and mold nutrition expert Dr. Andrew J. Moyer, fungal taxonomist Dr. Kenneth Raper, microbiologists Dorothy Fennell and Dr. L. J. Wickerham, and chemist Dr. Jacques L. Wachtel. Research began the day after the British scientists' visit, with one of them, Dr. Heatley, remaining a few months to help the team achieve its goal.[17]

Like other inventions, the process for mass producing penicillin evolved over time, discovery by discovery, with more than 500 experiments in the Peoria lab alone.[18] Never far from scientists' minds was what was at stake. By November 1941, just weeks before Pearl Harbor, Moyer

FIGURE 9
Dr. Andrew J. Moyer examines cultures for growth, Northern Regional Research Laboratory, Peoria, Illinois, 1940s. Collections of the USDA-NCAUR and Peoria Historical Society.

succeeded in increasing the penicillin yield substantially. He got there by making the critical discovery that the addition of the corn-steep liquor from Peoria breweries and distilleries produced ten times the previous yield. Replacing the sucrose used by the British scientists with lactose helped further, as did the addition of penicillin precursors, such as phenylacetic acid, to the medium (Figure 9).[19]

> It was stroke of good luck that the British scientists were directed to Peoria for it soon developed that, when a little lactose was mixed with the [corn-steep] liquor, there was an immediate tenfold increase in the yield of penicillin. It's not exaggeration to say that this discovery made commercial production eventually possible.[20]
> —W. H. Helfand, Merck Sharp & Dohme International

The story of how you found the organism discontented with the set-up and your improvement of its lot was what I would have expected; nevertheless the accomplishment is outstanding. Your long years of experience in making the "bugs" jump through the hoop were just what was needed.[21]

—Percy Wells, USDA, writing to Andrew Moyer

Nevertheless, by March 1942, the Peoria lab had produced only enough of the drug to treat a single case. Part of the problem was that the British penicillin strain had to be cultivated on the surface of nutrient mediums in wide-bottomed flasks. From their experience with deep fermentation techniques at the Color Lab and in Peoria, the scientists knew that to get commercial quantities of penicillin, the mold had to be grown while submerged in large, constantly agitated and aerated tanks. The British strain produced only traces of penicillin when grown this way.[22]

It was clear that a more productive strain of the *Penicillium* mold needed to be located—one that would flourish in the large drums required for commercial production. A worldwide search ensued. Moldy material, primarily fruit, was brought from as far away as India via the Army Transport command.[23] Team members also visited grocery stores and farmers markets looking for items molded by the *Penicillium* fungus. Using mold assays developed by Heatley, mycologists Raper and Fennell and a large group of assistants screened thousands of cultures for penicillin.[24] They worked seven days a week, ten hours a day.[25] Ironically, one of the best producers was found on a cantaloupe in a Peoria fruit stand.[26] Although the origin of the cantaloupe has been debated, its discovery was firmly rooted in the war-time civic spirit of Peoria. One story asserts that the moldy cantaloupe was brought in by a Peoria housewife who knew that the lab was looking for specimens. She is said to have dropped the cantaloupe off with lab guards without leaving her name. Another, more widely circulated story attributes the melon to lab technician Mary Hunt, who earned the nickname "Moldy Mary" as a result of her regular trips to scout for moldy produce (Figures 10 and 11).[27]

From a Spoiled Cantaloupe in Peoria . . .
the best of 100,000 strains of Penicillium

FIGURE 10
"From a Spoiled Cantaloupe in Peoria . . . the best of 100,000 strains of Penicillin," Abbott Laboratories Penicillin Products leaflet, 1948. Painting by Douglas Gorsline. Courtesy of the Abbott Historical Archives.

FIGURE 11
Lab technicians screening cultures for penicillin. Collections of the USDA-NCAUR and Peoria Historical Society.

Interestingly enough, the most important culture discovered was isolated from a moldy cantaloupe collected in Peoria.[28]

—Kenneth Raper, Northern Regional Research Lab

Our job not only was to find the right mold, but also to develop a medium which would assist the mold to produce penicillin.[29]

— Mary Hunt , Northern Regional Research Lab

Microbiologist Dorothy Fennell was the person who isolated the cantaloupe strain (Figure 12).[30] When submerged in vats of corn-steep liquor and pumped with air, it produced fifty times more penicillin than the original strain. Fennell would continue to play a significant role in the mass production of penicillin. Her role in the penicillin story, and that of other women in the international effort, is something that the Peoria *POI* team is working to document in greater detail to share through the project.

FIGURE 12

Microbiologist Dorothy Fennell examining spore structures. Northern Regional Research Laboratory, Peoria, Illinois, 1940s. Collections of the USDA-NCAUR and Peoria Historical Society.

Peoria's Northern Regional Research Laboratory continued to prototype procedures that achieved increasingly larger yields when tested in the lab's pilot plant facility. The lab's scientists shared their processes and samples of the new strain with other government, academic, and pharmaceutical labs furiously working to keep speed in the race against the war. The goal was to "obtain better penicillin-producing strains of mold, either by isolation from nature or by the production of mutations from previously known organisms by means of light, x-ray, or chemicals."[31] The Carnegie Institution irradiated the "Peoria cantaloupe" mold and produced an even higher producing variant, known as X-1612. Samples of the X-ray mutant strain were sent to the University of Minnesota and University of Wisconsin, where they were irradiated and made dramatically more productive.[32]

Penicillin production was quickly scaled up, and by 1943, the U.S. War Production Board was overseeing the large-scale production of penicillin at 21 pharmaceutical companies (Figure 13). Thanks to around the clock work, 2.3 million doses were available for the planned 6 June 1944 D-day invasion. It is estimated that penicillin saved the lives of 12%–15% of wounded Allied soldiers and increased the number of troops able to fight by treating other infectious diseases.[33] By 1945, penicillin was available without restriction to civilians and launched the antibiotic age. The Royal Society of Chemists and the American Chemical Society designated the USDA's Peoria lab an International Historical Chemical Landmark. Dr. Moyer was the first government scientist inducted into the National Inventors Hall of Fame.[34]

> The Peoria Facility was the only laboratory where the [use of] corn-steep liquor would have been discovered. We were lucky. So many things came together unexpectedly. It was the scientific problem of the century handed to us on a silver platter.[35]
>
> —Robert D. Coghill, Northern Regional Research Lab

ABOUT PENICILLIN

From Algiers comes this report by the Allied Medical Congress: "There never has been a therapeutic agent to compare with penicillin in its usefulness against a wide variety of diseases, including pneumonia, bone infections, syphilis, and a host of other infectious 'ills'."

More than *one hundred times as much penicillin* is being made today as there was a year ago—and the cost has been reduced 84 per cent. Remarkable changes have occurred in the method of manufacture since penicillin was first made at the Squibb Laboratories in 1940. Bottles once used for growing the mold have been replaced by huge tanks several stories high. Production time has been cut from two weeks to three days.

Military needs come first, but production is now great enough to provide limited amounts of penicillin for civilian use. The House of Squibb is proud to have shared in the development of this new medicinal agent that now is man's greatest defender against bacterial enemies.

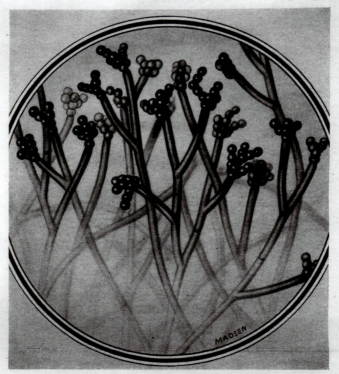

Through a microscope the fibres and spores of *Penicillium notatum* look like this. Growing in a liquid medium, this mold gives out golden droplets rich in penicillin—but the liquid must be concentrated over 30,000 times to obtain pure penicillin. The Squibb Laboratories were the first to obtain crystals of Penicillin Sodium from the natural product.

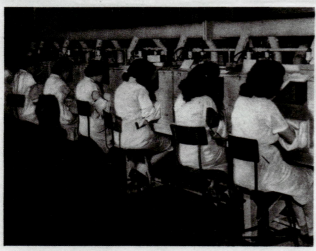

Unusual care maintains purity. Workers package penicillin in air-conditioned rooms, with ultraviolet lights to sterilize the air. For over two years, Squibb has been producing penicillin for the National Research Council and for the armed forces. Today, through designated hospitals, physicians can apply for the quantity of penicillin they need to treat infections.

New Squibb Penicillin Building. Giant tanks have replaced the glass bottles in which penicillin once grew so slowly. Instead of a few pounds, now over a ton of mold is grown each day, making possible a great increase in the production of penicillin.

SQUIBB
A name you can trust

FIGURE 13

Squibb Corporation penicillin advertisement, including workers packaging penicillin at the new Squibb penicillin building, 1944. Courtesy of the N. W. Ayer Advertising Agency Records, Archives Center, National Museum of American History, Smithsonian Institution.

PLACES OF INVENTiON PROJECTS AND PROGRAMS

By revisting the penicillin topic, the Peoria team had the opportunity to explore the subject more fully and to take advantage of resources and knowledge at all three institutions. The team has been successful in building upon the public's interest in the project and will continue to focus on Peoria as a place of invention in the years to come. Lectures and online exhibitions from the Peoria Historical Society will be a part of this exploration, as will a collaboration between USDA-NCAUR scientists and Smithsonian curators to identify and document women in science at the Northern Regional Research Laboratory (Figure 14).

Places of Invention Interactive Map Videos

Because of the international scope of the Peoria penicillin story, the team created four videos for the *POI* interactive map and left the door open to doing more. (One additional video might utilize oral histories of women who worked in the Peoria fermentation lab.) All videos will be incorporated as much as possible into current PRM exhibitions and web outlets.

Peoria, IL: A Place of Fermentation Innovation

This video summarizes why Peoria was the right place at the right time to develop innovations in the large-scale production of penicillin. An overview of Peoria as a place of invention is presented, including its history as a location where skilled people, abundant resources, and circumstance converged on the shores of the Illinois River. Local fermentation innovation is explored, including Peoria's history as a center of the nation's distilling and brewing industries. Viewers are encouraged to make connections between these factors and the Northern Regional Research Laboratory's leadership in fermentation research.[36]

Yellow Magic: British Discovery of Penicillin

The second video provides important historical context to the Peoria chapter of the international penicillin story. Sir Alexander Fleming's discovery of the mold's anti-bacterial characteristics and applications are described, along with advances made at Oxford University to find a way to create enough penicillin to save the lives of soldiers during World War II.[37]

Breakthroughs in Peoria at USDA's Northern Regional Research Laboratory

This video shows a place of invention and innovation in action and the chain of events that brought the penicillin challenge to Peoria. Viewers come to understand how the Peoria innovation story is one of expertise, risk-taking, collaboration, and luck. They learn how Dr. Andrew Moyer and his Peoria lab colleagues experimented to find improved culture mediums, processes for deep fermentation, and, by chance, a more-productive *Penicillium* strain discovered right in town.[38]

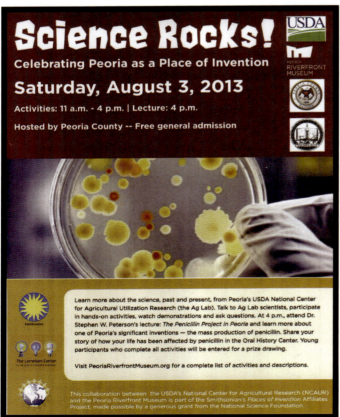

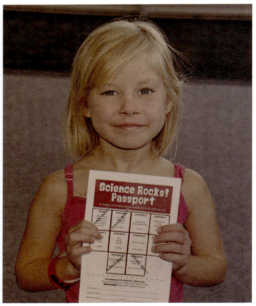

FIGURE 14

Scientists from the USDA-NCAUR leading the public in *Science Rocks!* hands-on activities, Peoria Riverfront Museum, Peoria, Illinois, 3 August 2013. Photographs by Christopher Coulter, Peoria Historical Society Collection, Bradley University Library. Flier image courtesy of the Peoria Riverfront Museum.

CONTRIBUTORS

Joyce Bedi is the senior historian at the Smithsonian's Lemelson Center for the Study of Invention and Innovation at the National Museum of American History. She is responsible for the center's scholarly publication program and website and assists with the development of scholarly programs and exhibitions. She is the coeditor, with Arthur Molella, of the Lemelson Center Studies in Invention and Innovation book series with MIT Press and of the first volume in that series, *Inventing for the Environment* (2003). Bedi has also authored publications and exhibitions on the work of Harold Edgerton in stroboscopic photography. Before joining the Lemelson Center in 1995, Bedi's research and curatorial career included positions at the MIT Museum, the Edison National Historic Site, the IEEE History Center, and the Powerhouse Museum in Sydney, Australia.

Laurel Fritzsch is a project curator on the *Places of Invention* project with the Lemelson Center for the Study of Invention and Innovation at the National Museum of American History (NMAH). She also cohosts and coproduces the Lemelson Center's podcast series and worked on the NMAH's exhibit *Making a Modern Museum: Celebrating the Fiftieth Anniversary of the National Museum of American History*. Fritzsch spent the previous five years as a museum curator at the Kansas Historical Society, where she helped develop three exhibitions, worked with the museum's collections, facilitated object donations, and worked with social media. She has also previously worked in the education division of the National Museum of Wales in the United Kingdom and as an archival assistant, assistant registrar, and archaeology assistant at the Wisconsin State Historical Society. Fritzsch has a B.A. in history from Lawrence University in Appleton, Wisconsin, and a M.A. in museum studies from the University of Leicester in Leicester, England.

John L. Gray is the Elizabeth MacMillan Director of the National Museum of American History. With a background as a banker and a belief in transforming museums through support of scholarship and education, Director Gray stakes out a clear vision of what's ahead for the National Museum of American History, the place where America comes to visit its rich and multilayered heritage. Prior to becoming the museum's ninth director, he was founding president of the Autry National Center of the American West, a successful merging of three cultural organizations: the Autry; Colorado's Women of the West Museum in Denver; and Los Angeles' oldest museum, the Southwest Museum of the American Indian. He also created the Institute for the Study of the American West, which supports the scholarly, interpretive, and educational activities of the Autry National Center. Commercial banking was the foundation of Gray's previous career. He served as executive vice president of First Interstate Bank of California in Los Angeles from 1987 until 1996 and worked at the Small Business Administration in Washington, D.C., for two years, 1997 to 1999. Gray has a bachelor's degree from C. W. Post College at Long Island University and a master's degree in business administration from the University of Colorado. He serves on the boards of the Global Center for Cultural Entrepreneurship in

Santa Fe and St. John's College in Annapolis and Santa Fe and received an honorary degree from Occidental College in Los Angeles.

Eric S. Hintz is a historian with the Lemelson Center for the Study of Invention and Innovation. He serves as a curator on the *Places of Invention* exhibition project and is responsible for producing the center's annual symposium series, "New Perspectives on Invention and Innovation." In addition, Hintz coordinates the Lemelson Center's fellowship and grant programs, assists in the collection of historically significant artifacts and documents, and pursues opportunities to speak and write about his scholarly interests. Hintz's research interests include the history of science and technology and U.S. business and economic history; he specializes in the history of invention and research and development (R&D). He has discussed his work on MSNBC and National Public Radio; his publications have appeared in the *Wall Street Journal*, *Technology and Culture*, the *Business History Review*, *Enterprise and Society*, and *Research-Technology Management*. Hintz is currently working on a book that considers the changing fortunes of American independent inventors from 1900 to 1950, an era of expanding corporate R&D. Hintz earned his B.S. in aerospace engineering from the University of Notre Dame (1996), then worked for nearly six years in San Francisco and Silicon Valley as a technology consultant for Accenture, a leading services firm. After leaving the corporate world, he taught both science and history at Sacred Heart Cathedral Preparatory High School in San Francisco before pursuing graduate studies at the University of Pennsylvania, where he completed his M.A. (2005) and Ph.D. (2010) in the history and sociology of science.

Anna Karvellas (coeditor) is the program specialist for the *Places of Invention* Affiliates Project, a new collaborative model for the Smithsonian, its Affiliate institutions, community partners, and the general public. She provides training and resources to help groups document their places of invention and discover how invention can be a transformative lens for understanding local history. Karvellas works with teams to develop public programming, oral history archives, and video content for the interactive map at the heart of the *Places of Invention* exhibition. Karvellas has nearly 25 years of experience editing complex, multiauthor print and online publications. Her research and curatorial work has ranged from William Steinway and nineteenth-century New York to regional American culture and identity as expressed through American music and decorative arts. Raised in Reston, Virginia, and a former longtime resident of New York City, she has a keen interest in urban planning and the relationship between space, community, and creativity. Karvellas joined the National Museum of American History in 2007 as managing editor of the William Steinway Diary Project in the Division of Culture and the Arts. She developed the William Steinway Diary website, the first publicly available online edition of the 2,500-page diary written by a key figure in the cultural, political, financial, and physical development of New York City. She also curated the related exhibition, *A Gateway to the 19th Century: The William Steinway Diary, 1861–1896*, in the Museum's Albert H. Small Documents Gallery. Prior to joining the Smithsonian, she worked as an editor, writer, and researcher for organizations such as W. W. Norton & Company, Time Inc., and Sotheby's. Karvellas is a graduate of the University of Michigan where she was the recipient of a 1992 Avery Hopwood and Jule Hopwood Prize: Special Award in Fiction.

Lorraine McConaghy is the project consultant for the *Places of Invention* Affiliates Project, which she directly inspired through her award-winning Nearby History work at the Museum of History & Industry (MOHAI) in Seattle, Washington. As MOHAI's Public Historian, she developed its core exhibition, *True Northwest: The Seattle Journey*, which was framed by four thematic through lines, one of which pursued metro Seattle's distinctive history of invention and creativity. With Dr. Margaret O'Mara, she also developed MOHAI's Bezos Center for Innovation, winner of a 2014 MUSE Award from the American Alliance of Museums. In addition to these duties, Dr. McConaghy curated Washington State Historical Society's *Civil War Pathways* exhibition and served as public historian for its Civil War Read-In. She is also a consultant on the Ford's Theatre Remembering Lincoln project. McConaghy completed her Ph.D. in U.S. history in 1993 at the University of Washington and has published articles and chapters as a public historian concerning readers' theater, oral history, and the Nearby History program. Her principal area of historical research is the antebellum and Civil War period in Washington Territory, particularly the U.S. Navy's Pacific Squadron. Her most recent book is *Free Boy: A Story* *of Slave and Master* (University of Washington Press, 2013), which concerns antebellum slavery in Washington Territory. Other recent publications include *New Land, North of the Columbia: Historic Documents That Tell the Story of Washington State from Territory to Today* (Sasquatch Books, 2011) and *Warship Under Sail* (University of Washington Press, 2009).

Arthur P. Molella (coeditor) is the Jerome and Dorothy Lemelson Director of the Smithsonian Institution's Lemelson Center for the Study of Invention and Innovation at the National Museum of American History. Under his direction, the center has developed a wide range of publications, exhibitions, and programs. He received his Ph.D. in the history of science from Cornell University. At the National Museum of American History, he has also served variously as curator of the Electricity Collections and chairman of the Department of the History of Science and Technology and of the Department of History. He was head curator of the Smithsonian's *Science in American Life* exhibition, cocurator of the international exhibition *Nobel Voices*, and most recently curator of *Making a Modern Museum: Celebrating the 50th Anniversary of the National Museum of American History.* He has published and lectured internationally on the relationship between science, technology, and culture and on museum exhibitions and strategies. His publications include volumes 1–4 of *The Papers of Joseph Henry* (coeditor), *Inventing for the Environment* (coeditor with Joyce Bedi, MIT Press, 2003) and *Invented Edens: Techno-cities of the 20th Century* (with Robert Kargon, MIT Press, 2008). Molella currently serves on the boards of the National Academy of Inventors and the MIT Museum. He was awarded a doctorate of science, honoris causa, from Westminster University in the United Kingdom and also serves as a senior lecturer in the Department of History of Science and Technology of Johns Hopkins University.

Robert E. Simon Jr. is founder of Reston, Virginia, one of the country's most famous planned communities and "new towns" of the 1960s. Reston—derived from Simon's initials, RES—was planned as an open community where people of all races, ages, and incomes could live, work, and play over the course of a lifetime. Key to this vision was the development of community and opportunities for people to gather in shared public spaces set within inspiring natural and architectural contexts. Simon was born in New York City and grew up on the Upper West Side. After graduating from Harvard University, he became head of his family real estate management and development business. It was the sale of Carnegie Hall, one of Simon's family properties, that allowed him to purchase 6,750 acres of rolling hills in Fairfax County, Virginia. He began development of Reston in 1961 with a diverse team of experts, including planners William Conklin and James Rossant. The first industry and residents moved to Reston in 1964, and in 2014 Reston celebrated its 50th anniversary—and Simon's 100th birthday. Today, Reston is a vital part of the Dulles Technology Corridor and home to over 60,000 residents and nearly as many workers. In 2002, the American Institute of Certified Planners presented Simon with its Planning Pioneer Award and named Reston a National Planning Landmark.

Monica M. Smith is the exhibition program manager for the Smithsonian's Lemelson Center for the Study of Invention and Innovation at the National Museum of American History. She serves as project director, principal investigator, and cocurator for the center's National Science Foundation–funded *Places of Invention* exhibition project. Previously, Smith was the project historian and then the second project director and principal investigator for the center's National Science Foundation–funded *Invention at Play* traveling exhibition, which won an AAM Excellence in Exhibition award and also a MUSE Gold Medal for its companion website. Other Smithsonian exhibition projects on which Smith has worked include *Making a Modern Museum: Celebrating the Fiftieth Anniversary of the National Museum of American History, Time and Navigation: The Untold Story of Getting from Here to There,* and *From Frying Pan to Flying V: The Rise of the Electric Guitar*. Smith's publications include "Invention at Play: An Award-Winning Interactive Traveling Exhibition" in *Museums at Play: Games, Interaction and Learning* (2011) and "The Electric Guitar: How We Got from Andres Segovia to Kurt Cobain" in *Regional Cultures in American Rock 'n' Roll: An Anthology* (2011), which was originally published in *American Heritage of Invention and Technology* (Summer 2004). She is also a featured speaker in the Smithsonian Channel's award-winning film *Electrified: The Guitar Revolution*. In addition to her tenure at the Lemelson Center since 1995, Smith served as editor in chief of the *Journal of Museum Education* from 2005 to 2008 and is on the board of directors of the Rotary Club of Washington, D.C. (2014–16).

INDEX

3M Company, 106
20th Century Fox, 164
21st Century Skills, 7

A

Abound Solar, Inc., 198
Adams, Alva B., 183
Advanced Micro Devices, 22, 39
Advanced Research Projects Agency
    (ARPA), 25, 199
Aetna Life Insurance Company, 110, 112
Affiliates Project Dispatches, 13, 223. *See*
    also penicillin, mass production
    of; Pittsburgh jazz; Seattle gam-
    ing industry
African American actors, 161
Agriculture Adjustment Act of 1938, 144
Albrecht, Bob, 29
Alcoa, Inc., 4
Alexander, Howie, 220
Allen, Paul, 72, 227
Altair 8800 computer, 29–30, 31, 32, 36, 38
Alto computer, 20, 27, 28, 33–34, 38, 39
Amelco Industries, 22
American Chemical Society, 149
American Mutoscope and Biograph
    Company, 159–160
American Publishing Company, 126
American Textile History Museum, 12
Apple, Inc., 1, 20, 31–35, 38–41
Apple I computer, 32–33
Apple II computer, 33
Apple Lisa computer, 33
Apple Macintosh computer, 20, 37–38
Armstrong, Louis, 216
ARPANet, 41
Ashby, Marty, 204–205, 217
Atkinson, Bill, 33

Atomic Energy Commission, 185
Augmentation Research Laboratory, 26
Austin, Chuck, 220
Autobee, Robert, 183
automobile manufacturing, 124–126

B

Bachmann, John, 111
Back to the Future summer camp, 225
Bailey, O. H., 118
Baker, Sandra, 217
Bakken, Connie Olson, 98, 99–100
Bakken, Earl, 9, 22, 89, 96–107, 109
Bakken Museum, The, 101
Ball, Joseph Arthur (J. A.), 168, 169, 178
Bambaataa, Afrika, 42, 50–52, 56, 58,
    63, 65
Bardeen, John, 22, 35
Barnard, Christiaan, 90, 108
Barrymore, John, 156
Base Hospital Number 26, 88
batteries, rechargeable, 192–194
Bauer, Tim, 188
Baxter, John, 178
b-boying, 45, 49–51
Beavers, Louise, 161
Becky Sharp (film), 173
Bell, Alexander Graham, 9
Bell, Charles, 220
Bell Labs, 4, 35
Benson, George, 203, 208, 214, 217, 220
Berkshire Historical Society, 12
Berkshire Museum, 12
Bettors, Harold, 220
Bezos Center for Innovation, 12, 68, 69,
    73, 80, 223, 225, 229
    audience engagement at, 225
    Lorraine McConaghy and, 223

    photographs of, 68–69, 73, 80
    video game exhibit at, 79–80
b-girling, 45, 49–51
bicycle manufacturing, 120, 122–124
Billings, Charles E., 117–118, 122
Billings & Spencer, 120
Billy Eckstine Orchestra, 219
Bilsback, Kelsey, 198
Biograph Company, 159–160
The Birth of a Nation (film), 159
Black, Samuel W., 203, 220, 221
Black Maria studio, 159
Blakey, Art, 214, 217, 219, 220
Blank, Julius, 22, 23
Bliss, Elijah Jr., 126
Block, Adriaen, 112
Blount, Ainsworth, 183
Blue Note Cafe, 210
Boeing, 226
Bogle, Donald, 161
Bohemian Foundation, 199
Bolden, Frank, 206
boom boxes, 63, 64
Boulder, Colorado, 185, 199, 200
Brand, Stewart, 26, 41
Branscomb, Lewis M, 185
Brattain, Walter, 22, 35
breakdancing, 49–50. *See also* b-boying;
    b-girling
breech-loading rifle, 119–121, 128
Brendle Group, 190, 191
bricolage, 9
Bronx, New York, 2, 42–65. *See also* hip-
    hop music
    compared with Silicon Valley, 8
    cultural diversity in, 52, 64
    economic troubles in, 42–44
    history of, 43–45

overview, 42
revitalization of, 64–65
Bronx River Houses, 50
Brooks, Cecil, 214
Brown, Chuck, 55
Brown, James, 52
Brown, Leroy, 212, 213–214
Brown, Ray, 208, 213, 214, 217, 220
Brown Derby restaurant, 5, 162, 163–164
Brundage, Jennifer, 205
Brunner, Matt, 73
Buchanan, Ruby Young, 212
Buhl Foundation, 221
Bullitt Foundation, 225
Bush, Vannevar, 89
Bushnell, David, 113
Busicom corporation, 36
Busy Bee (MC), 60, 61

C

Cache de la Poudre River, 182
Cambers, Paul, 220
Campbell, Clive. See Kool Herc
Campbell, Tony, 220
Cardiac Pacemakers, Inc. (CPI), 107
cardiac surgery, 90–91, 92–96
Carnegie Institution, 149
cartoons, 172–173
Casablanca (film), 160
case studies. See also Bronx, New York;
        Fort Collins, Colorado; Hartford,
        Connecticut; Hollywood, Cali-
        fornia; Medical Alley, Minnesota;
        Silicon Valley, California
    arrangement in exhibit, 1–3
    crowdsourced, 4, 11
    list of, 2
    selection of, 7
Central Radio Propagation Laboratory,
        185
Century 21, 227–228
Chaplin, Charlie, 160
Chardack, William, 104
Chardack-Greatbatch pacemaker, 105
Charter Oak Hall, 116, 126
Chase, Charlie (DJ), 50, 64
Cheney Brothers silk mills, 119
Cherian, Sunil, 195–197
chord keyset, 36
Civic Arena, 217
Clarke, Kenny, 214, 217, 220
clean energy. See energy technology
        innovations
Clemens, Samuel L. See Twain, Mark
Clemetson, Lynette, 217
Cobain, Kurt, 226
Cobb, Robert, 163

Cobb, Sally Wright, 163
Coghill, Robert D., 139, 145, 149
Coles, Charles "Honi," 216
Collins, W. O., 180
Colorado. See Fort Collins, Colorado
Colorado Agricultural College, 182, 186
Colorado-Big Thompson (CBT) project,
        183–184
Colorado Clean Energy Cluster, 189, 190
Colorado State College of Agriculture
        and Mechanical Arts, 186
Colorado State University (CSU), 185, 186
Colorado State University Engines and
        Energy Conversion Laboratory
        (EECL), 183, 187–189, 196, 197,
        199, 200
color films. See Technicolor technology
Colt, Elizabeth, 118
Colt, Samuel, 10, 19, 22, 114–117, 126,
        129, 132
Colt Armory, 110, 115–119, 126, 128, 130,
        132, 134
Colt's Armory Band, 117, 126, 127
Columbia brand bicycles, 122, 123, 124
computers. See personal computing;
        Silicon Valley, California
Comstock, Daniel Frost, 164–168, 178
Condon, Tom, 134
Connecticut River, 110–113, 115, 118, 132
Conner Prairie Interactive History
        Park, 13
Contagious Creativity, 72, 77
Control Data Corporation, 107
Conzo, Joe, 45
CooperSmith's Pub & Brewing, 199
Corning corporation, 4
Costa, Johnny, 214, 220
Count Basie, 219
Cox, Etta, 220
Coyle, Bud, 22
Crawford, Ray, 213
Crawford Grill nightclub, 203
Cray, Seymour, 107
Cray Research, Inc., 107
creative class, 5, 9, 197
Crosby, Bing, 158
CSU Ventures, 199
Cuddihy, Charlie, 107
"Cultures of Innovation" conference, 6
Culver, Lawrence, 158
Curtiz, Michael, 160

D

Davey, Ken, 46
da Vinci, Leonardo, 5
Davis, Billy "Stinky," 213
Davis, Nathan, 220

Dawson, Mary Cardwell, 208
Deal, Jim, 73
Dee, Mary, 212
de Forest, Lee, 172
de Portolá, Gaspar, 15–16
Department of Defense, 4, 18–19, 25
Department of Energy's (DOE) national
        labs, 185
DeSena, Alphonse, 12
Detre Library and Archives, 204, 221
DeWall, Richard, 94, 95, 96
DeWall-Lilllehei bubble oxygenator,
        94, 95
Dickson, William Kennedy Laurie, 159
Disco Wiz (DJ), 46, 47, 50, 62, 63
Disney, Roy, 173
Disney, Walt, 163, 171, 172–173, 178
Dispatch videos and programs
    After-School Video Game Design
            Club (program), 84
    Breakthroughs in Peoria at USDA's
            Northern Regional Research
            Laboratory (video), 151
    Building Bridges in Music Education
            (web portal), 221
    "A Byte of Seattle: The Rise of
            Seattle's Gaming Industry"
            (event), 80–81
    The Crawford Grill (video), 217, 220
    Gaming in Greater Seattle (video),
            79–80
    Past, Present, and Future of Video
            Games (exhibition), 80
    Peoria, IL: A Place of Fermentation
            Innovation (video), 151
    Pittsburgh Jazz Innovation (1970–20)
            (video), 217
    "Places of Invention: Pittsburgh Jazz
            Innovation (1970–20)" (program),
            220
    Scaling Up: Peoria and the Mass Pro-
            duction of Penicillin (video), 153
    Science Rocks! (program), 152–155
    "Seattle Anti-Freeze: Games"
            (event), 81, 84
    The Sounds of Pittsburgh Jazz (video),
            219–220
    Yellow Magic: British Discovery of
            Penicillin (video), 151
Divjak, Helen, 70
DJ Baron, 49
DJs in Bronx. See hip-hop music
Dompier, Steve, 31
Donaldson, Cheryl, 199
Dorsey, Judy, 190
Dow Chemical, 185
Dowe, Al, 220

Dunlap, Anna Simmons "Birdie," 211
Dunlap, Shine, 211
DuPont corporation, 4

E

Eames, Hayden, 122, 124, 125
Eckstine, Billy, 204, 205, 215, 219, 220, 221
Edison, Thomas, 5, 9, 159, 222
Elbert, Samuel, 182
Eldridge, Roy, 216, 218, 220
E-Line Media, 84–85, 225
Ellington, Duke, 216, 219
Elwell, Cyril, 16
energy technology innovations. *See also*
    Fort Collins, Colorado
    Colorado Clean Energy Cluster,
      189–190
    cookstoves, 189
    FortZED and, 190, 197, 199
    New Belgium Brewing Company
      and, 190–192
    power alternatives, 194–195
    rechargeable batteries, 192–194
    research on, 186–189
    smart grids, 189, 195–196
    start-ups, 189–190
Engelbart, Douglas, 25–27, 34, 36, 38–39
Engineering Research Associates, 107
Engines and Energy Conversion Labo-
    ratory (EECL), 187, 188–189, 196,
    197, 198, 199, 200
English, Bill, 26, 38
Envirofit International, 188, 189–190
Ethernet, 16, 27, 39

F

Factor, Frank (Max Factor Jr.), 175
Factor, Max, 174–176
Fairbanks, Douglas, 160
Fairchild Semiconductor, 20, 22, 23, 37
Fairfield, George, 119, 122
Fantasia (film), 178
Federal Telegraph Company, 16–17
Feld, Brad, 200
Feldman, Maryann, 5, 8, 201
Felsenstein, Lee, 30, 31
Fennell, Dorothy, 145, 147, 148, 149
fermentation innovation, 142, 151. *See*
    also penicillin, mass production of
Fermi National Accelerator Laboratory
    (Fermilab), 4
Fields, W.C., 163
film industry. *See* Hollywood, California
firearm manufacturing, 115, 128, 132. *See*
    also Colt Armory
Fleming, Alexander, 139, 151

Florey, Howard, 144–145
Florida, Richard, 5, 200, 201
Flowers and Trees (cartoon), 173
Fort Collins, Colorado, 180–201. *See also*
    energy technology innovations
    in 1800s, 180–183
    in 1950s, 184–185
    in 1990s, 185–188, 190
    in 2000s, 188–197
    in 2010s, 188–189, 197–201
    agriculture in, 182–183
    vs. Boulder, Colorado, 200
    as educational center, 182–183
    factors contributing to innovation in,
      200–201
    federal government presence in, 185
    FortZED district in, 190, 199
    funding in, 197, 199
    manufacturing in, 185–186
    New Belgium Brewing Company in,
      190–192
    research in non-energy-related
      fields, 197
    technology incubators in, 199
    water supply for, 183–185
    zero-energy district in, 190, 199
FortZED, 190, 197, 199
Forward Thrust program, 226
Fox, William, 160
Frankenstein (film), 97, 98
French, Gordon, 29, 31
Fries, Ed, 72, 73, 74, 75, 77, 78, 79–80, 81
Fulton (steamboat), 113

G

Gage, Andrew, 104
Gaiser, Megan, 72, 77–78, 79
gaming industry. *See* Seattle gaming
    industry
gangs, 44–45, 50–51
Garfield, Richard, 72, 74, 75, 78, 79
Garland, Harry, 31
Garner, Erroll, 204, 208, 214
Garrity, William, 172
Gates, Bill, 72, 227
GeekWire, 81, 225
Geiselman, Annette, 199
Gendron, Tom, 195
General Electric, 18
General Hospital Number 26, 88, 91
George, Nelson, 60
Gertner, Jon, 4
Gibbon, John, 93
Gibbon's Model II oxygenator, 93
GI Bill, 88, 98, 134
Gillespie, Dizzy, 216

Global Innovation Center for Energy,
    Environment, and Health, 189
G Man, 60
Goldwyn, Samuel, 160
Gone with the Wind (film), 161, 178
Good Roads movement, 122, 123
Goodyear, Charles, 113
Google, 1, 41, 75
Gott, Vincent, 94, 95–96
Govanucci, Renée, 204–205, 221
Graham, Margaret, 4
GrandMaster Caz, 45, 64, 65
Grandmaster Flash, 8, 9, 42, 51, 52,
    53–56, 58, 59, 60, 61, 62
Grandmixer D.ST, 64, 65
GrandWizzard Theodore, 42, 45, 52,
    56–58, 62, 65
Grant, Ellsworth, 113
graphical user interface (GUI), 26, 27,
    33–34, 35, 36, 38, 39
Grasso, Francis (DJ), 45
Greatbatch, Wilson, 104–105, 106, 107
Greater Seattle, Washington. *See* Seattle
    gaming industry
Griffith, Andy, 52
Griffith, David Wark (D. W.), 159–160
Grinich, Victor, 22, 23
Grove, Andy, 24
The Gulf Between (film), 166–167, 176
gun manufacturing, 114–115, 116, 117,
    120. 128, 131. *See also* Colt Armory

H

Haines, Richard, 170–171
Hall, John, 120
Hampton, Slide, 215, 220
Hancock, Herbie, 64
Hansen, Charles, 183
Hanson, Roger, 102
Harper, Nate, 213
Harper, Walt, 208, 213, 220
Harris, Charles "Teenie," 217, 220
Harris, Joe, 213, 220
Harrison, Nelson, 203, 217, 220, 221
Hartford, Connecticut, 110–135
    in 1800s, 110–132
    in 1900s, 133–134
    in 1990s, 144
    abandonment of factories in, 133–134
    Colt Armory in, 110, 114–117, 126
    Connecticut River and, 111–112
    decline of, 134–135
    development of manufacturing pro-
      cesses in, 127–132
    early inventors in, 114–127
    factory laborers in, 126

government/military funding in, 128
insurance industry in, 112–113
municipal politics of, 133
people in history of, 114–127
revitalization of, 134
steam power revolution's impact on, 113
suburban migration from, 133–134
transformation to manufacturing hub, 113
Hartford Bank, 113
Hartford Machine Screw Company, 120, 122
heart-lung machine, 93–96
heart valve, 107
Heatley, Norman, 144, 145, 147
Helfand, W. H., 146
helical reservoir oxygenator, 94, 95
Her Interactive, 72, 77
Hermundslie, Elaine Olson, 98, 100
Hermundslie, Palmer, 98–99, 105, 106
Hertzfeld, Andy, 33
Hewlett, William R. "Bill," 17
Hewlett-Packard Company, 17, 19, 20, 23, 32, 39
Hill District neighborhood, 204, 205, 206, 208, 210, 212, 217, 218, 220, 221
Hines, Baby, 210
Hines, Earl "Fatha," 204, 215, 216, 220
hip-hop music. See also Bronx, New York
    breakdancing to, 49–50. See also b-boying; b-girling
    contests, 62–63
    disco music and, 45–46
    gangs and, 50–51
    Jamaican community and, 45–49
    Latino community and, 49–50
    obscure records played with, 51–52
    peek-a-boo system, 55–59
    performance venues for, 58–63
    popularization of term, 51
    quick mixing in, 55
    scratching in, 56–58
    sound systems for playing, 42, 45–49, 52–53, 58
    spread beyond Bronx, 63–65
    techniques used in, 45–46, 49–50, 54–58
    toasting in, 49
    turntable use in, 54–58
    women in, 45
    Zulu Nation and, 50–51
Hoddeson, Lillian, 4
Hoefler, Don, 20–21, 22
Hoerni, Jean, 22, 23
Hoff, Marcian "Ted," 36

Hoffman, Joanna, 33, 40
Holkins, Jerry, 72, 75–76, 78, 79
Holland, Dan, 204, 205, 221
Holland, John, 70
Hollywood, California, 156–179. See also Technicolor technology
    in early 1900s, 156–160, 165–168
    in 1920s, 157, 158, 159, 172
    in 1930s, 157–166, 169–179
    in 1940s, 157, 161
    actors in, 161
    extras in, 161, 162
    hierarchy in, 161–164
    Hollywoodland sign, 157
    naming of, 156
    producers in, 160–161, 162
    wages in, 161–162
    World War II and, 179
Homebrew Computer Club, 20, 29, 30–32, 33, 36, 38, 40
Honeywell corporation, 107
Hooker, Thomas, 112
Hopkins, Miriam, 173
Hopper, Hedda, 164
horseless carriage, 124–125, 133
Horsetooth Reservoir, 184
"Hot Spots of Innovation" symposium, 7
Hounshell, David, 4
Howard, Brian, 33
Howard County Historical Society, 13
Howe, Elias, 113
Hudnut, Paul, 188, 189
Hudson, George, 213, 214
Hudson, Kristina Erickson, 76–77
Humphries, Roger, 217, 220
Hunt, Marsha, 175
Hunt, Mary, 147, 148
Hunter, Samuel, 103
Hunter-Roth bipolar platform electrode, 103
Hurricane Club, 211
Huys de Goede Hoop, 112
Hyman, Phyllis, 220

I
icons, 34, 38
IDEO design firm, 9
Illinois River, 137. See also penicillin, mass production of; Peoria, Illinois; Peoria Riverfront Museum (PRM)
    flooding of, 138
    rise as center of agriculture, 141–144
Illinois River Encounter exhibition, 138
Imagine Cup, 225
Imitation of Life (film), 161
Indiana Gas Boom Heritage Area, 13

Innoskate, 225
innovation
    critical analysis of, 225
    distinction from invention, 7
    history as lens of, 223
    importance of community in, 10, 224
    leading to more innovation, 224
    learning about through history, 2–4
    museums as essential to, 13, 223, 228
inside contracting, 117
insurance industry in Hartford, 112–113
integrated circuits, 36–38
InteGrid Lab, 196–197
Intel Corporation, 5, 20, 23–25, 36, 38
Internet, 41
Intersil corporation, 22
In the Chips board game, 40
invention
    distinction from innovation, 7
    museums essential to, 13
"The Inventor and the Innovative Society" symposium, 5
inventors, public perception of, 222
iTunes, 41
Iwerks, Ub, 172

J
Jackson, John, 219
Jackson, Wilfred, 172
Jaffa, Henri, 177
Jamaican community, 45–49
Jamal, Ahmad, 203, 208, 209, 212, 214, 216, 217, 219, 220
jazz music. See Pittsburgh jazz
The Jazz Singer (film), 171
Jazzy Jay (DJ), 65
Jefferson, Eddie, 220
Jobs, Steve, 9, 31, 32, 33, 38, 41
Johnson, Andrew, 180
Johnson, James Jr., 220
Johnson, Sam, 213–214
Johnson, Steven, 201
Joint Institute for Laboratory Astrophysics (JILA), 185
Jones, Robert Edmond, 167, 177
Jones, Sean, 220
Jones, Steve, 199
Jordan, David Starr, 16
Jordan, Kim, 190–191
Josie, Marva, 220
Judith of Bethulia (film), 159

K
Kalb, Jeffrey, 25
Kalmus, Herbert, 164–165, 166, 167, 168, 171, 172, 173, 176, 178

Kalmus, Natalie (Dunphy/Dunfee), 176–177, 178
Kankakee Torrent, 138
Kare, Susan, 33–34, 35, 38
Karloff, Boris, 97
Katz, Mark, 55
Kay, Alan, 26, 28, 34, 38
K-Dets band, 209
Keliher, Alice, 161
Kelley, David, 9
Kennedy, Joe, 213
keyboards (computer), 30, 32, 36, 38
Kimery, Ken, 205, 216
King, Calvin, 212
Kips Bay Medical, 107
Kirklin, John, 93
Kleiner, Eugene, 22, 23
Kodak, 20
Kolb, Adrienne, 4
Kool DJ AJ, 51, 52, 63
Kool DJ Red Alert, 65
Kool Herc (DJ), 42, 46–49, 50, 51, 52, 53, 56, 58, 60, 61, 62
Korda, Alexander, 160
Krahulik, Mike, 72, 75–76, 78, 79

L

La Cucaracha (film), 173
Lakeview Museum of Arts and Science, 137–138
La Rock, Coke (MC), 49
Last, Jay, 22, 23
Latino community, hip-hop music and, 49–50
Lawrence, Jacob, 226
Lawrence, Richard S., 121
League of American Wheelmen, 122
Lebesch, Jeff, 190
Lécuyer, Christophe, 4
Leftwich, Mariruth, 207
Lemelson Center for the Study of Invention and Innovation, National Museum of American History, 1–12, 203, 205, 223, 225
Lemelson Institute on Places and Invention, 6–7
LeRoy, Mervyn, 177
Leslie, Stuart W., 4, 5, 8, 18
Lewis, F. John, 91
LifeScience Alley, 108–109
Light, Jennifer, 4
Lillehei, C. Walton, 90, 91–96, 97, 100, 101–102, 104, 106, 107, 108, 109
Lillehei, Katherine, 92–93
Lillehei, Richard, 107
Lincoln, Abraham, 120, 180

Lincoln, George S., 129
Lincoln Milling Machine, 116, 129
Lisa computer, 33
Litton, Charles, 17, 19
Litton Engineering, 17
Litton Industries, 17, 23, 39
Livingston, Theodore. *See* Grand-Wizzard Theodore
Lockheed Missiles and Space, 18
Lorenz, Nathan, 188
Lowell Telecommunications Corporation, 12
Ludwig, Gene, 220
Lukaszewicz, Kate, 205, 207, 220

M

Macintosh computer, 20, 33–34
Majestic Pictures, 159
Mamoulian, Rouben, 173, 177, 179
Manchester Craftsmen's Guild, 204
Mancini, Henry, 214, 220
Mann, Louise, 210
Markusen, Ann, 5, 10
Marmarosa, Dodo, 220
Marsh, Bob, 31
Marx, Groucho, 163
Maryland Technology Development Center, 6
Mauston, Warren, 103
Maxim, Hiram Percy, 124–126, 133
Mayer, Louis B., 160
Mayo, Charles, 88
Mayo, William, 88
Mayo Clinic, 86, 88–89, 90, 93–94, 106, 109
Mayo-Gibbon heart-lung machine, 93–94
McCauley, Tara, 85
McClain, J. C., 212
McDaniel, Hattie, 161
MCG Jazz, 12, 204, 205, 217, 218, 221
McVickers, Carl, 208
McWilliams, Carey, 161
McWilliams, Carl, 185–186
McWilliams, Karen, 185–186
Mean Gene (DJ), 56, 60
mechanical heart valve, 107
Medical Alley, Minnesota, 86–109
    in 1800s, 86–88
    in 1940s, 88, 98–100
    in 1950s, 90–97, 100–105
    in 1960s, 105–106, 107
    in 1970s, 107–108
    in 1980s, 108–109
    in 2000s, 109
    cardiac surgery in, 90, 92–97
    funding of, 108–109
    overview, 86

pacemaker invention in, 100–106
people in history of, 89–101, 107–108
University of Minnesota and, 87–91
Medtronic, Inc., 9, 96–100, 101, 102, 103–106, 107, 109
Melen, Roger, 31
Mercedes Ladies DJ Baby D, 58
Merrell, Alton, 220
Metcalfe, Robert, 16
MGM Studios, 160, 162, 164
microchips, 35, 36,
Micro Instrumentation Telemetry Systems (MITS), 29, 30, 38
Microsoft, 11, 67, 72–74, 75, 76, 78, 225, 226, 227
microwave industry, 17
military influence
    in Hartford, 128
    in Silicon Valley, 17–20, 27, 39
Minneapolis, Minnesota. *See* Medical Alley
Mitchell, Grover, 208, 220
MITS Altair 8800 computer. *See* Altair 8800 computer
MNBIO, 108
Moore, Fred, 29, 30
Moore, Gordon, 22, 23, 24, 36
Moore's Law, 36, 39
mouse (computer), 20, 26, 27, 33, 34, 35, 38, 39
movie industry. *See* Hollywood, California
Moyer, Andrew J., 145–147, 149, 151
Mozart, Wolfgang Amadeus, 219
Muir, John, 17
Murray, Albert, 214
Muse, Clarence, 161
Museum of History and Industry (MOHAI), 67–85, 223, 225. *See also* Bezos Center for Innovation; Seattle gaming industry
museums
    engagement with community, 223–224
    as essential to innovation, 13, 222, 224
    innovative techniques in, 225
Musicians Club of Local 471, 211–212
Mutoscope, 159

N

Nash, Ogden, 90
National Bureau of Standards, 185
National Center for Agricultural Utilization Research (USDA-NCAUR), 139, 141, 151, 153, 155

National Center for Atmospheric Research, 185
National Museum of American History, vii. *See also Places of Invention* exhibit
National Oceanic and Atmospheric Administration's Earth System Research Laboratory, 185
National Science Foundation, 7, 10
National Semiconductor, 22, 39
National World War II Museum, 13
Nearby History program, 11–12
Negri, Joe, 214, 220
Netflix, 41
New Belgium Brewing Company, 5, 190–192, 199
Newell, Gabe, 73
nightclubs, 203, 208
Norris, William, 107
Northern Colorado Clean Energy Cluster, 189
Northern Colorado Water Users Association, 183
Northern Regional Research Laboratory, 139, 144, 145, 146, 148, 149, 151, 153
Noyce, Robert, 5, 22, 23, 24, 25, 36

O

Oasis restaurant, 30
Ogg, Bryan J., 144
Ohio Center for History, Art, and Technology, 12
Olson, Connie, 98
Olson, Elaine, 98
O'Mara, Margaret, 223
On With the Show (film), 171
Orwell, George, 34
Owens Corning, 12

P

Palo Alto Research Center (PARC), 26–28, 33, 34, 38, 39
pacemaker, invention of, 100–106
Pacific Science Center, 225
Packard, David, 17, 19
Paige, James, 127
Paramount Studios, 160, 164
Parker, Charlie, 9
Parker, Leo, 219
Parlan, Horace, 220
Parsons, Louella, 163
Partnership for 21st Century Skills, 7
peek-a-boo system, 52–56
penicillin, mass production of, 136–150. *See also* Illinois River;

Peoria, Illinois; Peoria Riverfront Museum (PRM)
  discovery of yield increase method, 145–148
  interactive map video about, 151
  lab opened at Peoria, 144–145
  overview, 139
  women in, 147–148
  World War II and, 139, 140, 145, 149
  X-1612 strain, 149
Penny Arcade, 72, 75–76
Penny Arcade Expo (PAX), 72, 81, 82–83
People's Computer Company, 29
Peoria, Illinois, 136–155. *See also* dispatch videos and programs; penicillin, mass production of
  as brewery and distillery town, 142–144
  effect of Great Depression and prohibition on, 144
  fermentation innovation at, 142
  lab opened at, 144–145
  rise as center of agriculture, 141–142
Peoria Historical Society, 12, 141, 143, 151, 153
Peoria Riverfront Museum (PRM), 12, 137–139, 152–155. *See also* penicillin, mass production of; *Science Rocks!* program
personal computing. *See also* Apple, Inc.; Silicon Valley, California
  Altair 8800 computer, 29–30, 32, 36, 38
  Alto computer, 20, 27, 28, 33–34, 38, 39
  Apple I computer, 32–33
  Apple II computer, 33
  Apple Lisa computer, 33
  Apple Macintosh computer, 20, 37–38
  graphical user interface (GUI), 26, 27, 33–34, 35, 36, 38, 39
  Homebrew Computer Club, 20, 29, 30–32, 33, 36, 38, 40
  icons, 34, 38
  integrated circuits for, 20, 25, 35–36
  keyboards, 30, 32, 36, 38
  microchips, 35, 36
  mouse, 20, 26, 27, 33, 34, 35, 38, 39
  Stanford University's influence on, 14, 16, 17, 19, 20, 39
  transistors and, 35–36
  Xerox Star computer, 33
Perti, Andrew, 81
Peterson, Stephen W., 153
Phoenix Iron Works, 129
Picker International, 107
Pickford, Mary, 160
Pinhead Institute, 13

Pittsburgh jazz, 202–221. *See also* Senator John Heinz History Center
  defining sound of, 214–217, 218
  interactive map videos about, 218–220, 224
  music education and, 207–209
  nightclubs featuring, 209–214
  outreach program for, 217
  research of, 203–205
Pittsburgh, Pennsylvania, 202–221. *See also* dispatch videos and programs; Pittsburgh jazz
Pittsburgh Public Schools System, 207–209
Places of Invention Affiliates Project, 10–13. *See also* Affiliates Project Dispatches
Places of Invention exhibit, vii
  floor plan, 3
Places of Invention map, discussion of, 10–13
"Places of Invention" workshop, 6
Pogany, Willy, 171
Ponder, Jimmy, 220
Pope, Albert A., 122–124, 125, 126, 128
Pope Manufacturing Company, 122–125
Popenoe, Chuck, 6
Popmaster Fabel, 50
Porter, Michael, 4, 201
Powerhouse Energy Campus, 198, 199
Power of Play event, 77
Powers Cinephone, 172
Pratt, Francis, 119, 129
Pratt & Whitney Aviation, 133
Pratt & Whitney Machine Tool, 119, 121, 125, 133
Prieto, Amy, 192–194, 195, 197–198, 199, 200
Prieto Battery, 192–194, 199, 200
Processor Technology Corporation, 31

Q

quick mixing, 52–56, 58
QWERTY keyboard, 36, 38

R

Raft, George, 163
Rane Empath DJ mixer, 59
Raper, Kenneth, 145, 147, 148
Raskin, Jef, 33
REACTOR group, 77
Reagan, Neil, 163
Reagan, Ronald, 91, 163
rechargeable batteries, 192–194
Reclamation Act of 1902, 183
Reeser, Tim, 199

Regional Advantage (Saxenian), 5
regionalism, as corporate strategy, 4
Reliance production company, 159
Renewable and Distributed Systems
    Integration (RDSI), 197, 199
Rentschler, Frederick, 133
Rescued from an Eagle's Nest (film), 159
Reston, Virginia, v–vi
revolvers, Colt, 114–117
Rickey's Restaurant, 25
Ritchie, T., 66
RKO Pictures, 164
Robbins, Samuel, 121
Robbins & Lawrence, 121
Roberts, C. Sheldon, 22, 23
Robinson, Prince, 210–211
Rock, Arthur, 22, 23
Rocky Mountain Innosphere incubator,
    197, 199
Roe, Joseph Wickham, 115
Rogers, Will, 163
Root, Elisha K., 115, 117, 119, 127, 129–131
Rosten, Leo C., 161
Roth, Norman, 102
Roundhouse restaurant, 25
Royal Society of Chemists, 149
Royal Typewriter Company, 133

S

Saddler, Joseph. *See* Grandmaster Flash
Saint Mary's Hospital, 95–96
Saint Paul, Minnesota, 88. *See also*
    Medical Alley
Sanborn Company, 100
San Francisco, California, 16
Santa Clara Valley, California, 14–16, 17.
    See also Silicon Valley, California
    in 1700s and 1800s, 15–16
    in mid-1900s, 17–20
    climate of, 14–15
    establishment of Stanford University
        in, 16
    military and government presence
        in, 18–19
    photograph of, 17
    role in cultivation of electronics
        industry, 16–17
    transformation to Silicon Valley,
        20–24
Saxenian, AnnaLee, 5, 10, 201
Schenck, Joseph, 160, 164
Schureman, Kate Neumiller, 141, 153
Science, the Endless Frontier (Bush), 89
ScienceWorks Hands-on Museum, 13
scratching, 56–58
screw-making machine, 120, 122

Seattle, Washington, 226–227. *See also*
    dispatch videos and programs
Seattle gaming industry, 66–85
    After-School Video Game Design
        Club, 84–85
    community in, 78–80
    companies in, 72
    history of, panel and presentation
        on, 83–84
    interactive map video, 82
    interviews about, 73
    liquidity of, 76–78
    mentorship programs in, 78–79
    Microsoft and, 72–76
    Museum of History & Industry and,
        66–85
    Penny Arcade Expo (PAX), 72, 81,
        82–83
    statistics about, 71
    women in, 79
Seattle Journey exhibit, 223, 225
Senator John Heinz History Center, 12,
    203–205
    archival resources, 220–221
    interactive map videos, 217–219
    jazz exhibition at, 203
    outreach program, 217
    Pittsburgh: A Tradition of Innovation
        exhibition, 203, 220–221
Serra, Junípero, 16
Shapiro, Morse, 90
Sha-Rock (MC), 51
Sharps, Christian, 120–122, 128, 129
sheet bubble oxygenator, 94
Shockley, Mary Bradford, 22
Shockley, William, 20–21, 22, 23, 25, 36, 40
Shockley Semiconductor, 20–22, 40
Shumway, Norman, 90
Sidney, George, 160, 162
Silicon Valley, California, 14–41. *See also*
    personal computing; Santa Clara
    Valley, California
    in 1970s, 21, 27–34
    in 1980s, 23–24, 33–36
    in 1990s, 42
    compared with Bronx, New York, 8
    in current times, 40–41
    efforts to replicate, 8
    emergence of, 20–24
    information flow in, 25
    intra-company communication in, 25
    gender imbalance in, 40
    low-skilled work at, 41
    microchip invention in, 22
    military and government influence
        on, 18–19, 26

overview, 14
    poster map of, 15
    tools and parts used by inventors
        in, 9
    transistor invention in, 22
    work culture in, 23–25
    World War II and, 16, 18, 25, 39
Silk, Van, 51
Silly Symphony cartoon series, 173
Simon, Robert E. Jr., v–vi, 10
Sisco, Richard, 50
Sloan Automotive Laboratory, 198
Sloat, John D., 16
SmartBolt, 6
smart grid technology, 195–196
Smith, Burrell, 33
Smith, John K., 4
Smith, Sandra, 205
Smithsonian Affiliations, 10, 12, 205
Smithsonian Folkways recordings, 208
Smithsonian Jazz Oral History Project,
    205, 216
Snow White and the Seven Dwarfs
    (cartoon), 173
Society for Savings, 113
Sol-20 microcomputer, 31
solid-state transistor, invention of, 21, 22
Solix BioSystems, 190
Soper, Taylor, 81
sound systems, 42, 58. *See also* Kool Herc
    for disco, 45–46
    Grandmaster Flash's influence on,
        52–55
    Jamaican community and, 46–49
    peek-a-boo system, 55–56
    salvaged parts used for, 8–9
Souther, Henry, 122
Southern Oregon Historical Society, 13
Sparks, Melvin, 52
Spaulding, George "Duke," 212, 220
Spencer, Christopher M., 119–120, 122
Spencer Repeating Rifle Company,
    119–120
Spirae, Inc., 189, 195–196, 199, 200
Squibb Corporation, 150
Stanford, Jane Lathrop, 16
Stanford, Leland, 16
Stanford, Leland Jr., 16
Stanford Industrial Park, 19–20
Stanford Research Institute (SRI), 19,
    25–26, 38, 39, 41
Stanford University
    cooperation with industry, 19–20
    establishment of, 16
    influence on personal computing,
        20, 39

military and government patronage, 19
Starski, Lovebug, 51
Staton, Dakota, 215, 220
steam power revolution, 113
Steffes, Al W., 90
Steinel, Alvin, 183
Stitt, Sonny, 219
St. Jude Medical, 107
Stowe, Harriet Beecher, 126
Stravinsky, Igor, 9
Strayhorn, Billy, 203, 205, 208, 214, 216, 217, 220
The Street exhibition, 137, 139, 141
Stress Indicators, Inc., 6
Strong, Tim, 220
Sugar Hill Gang, 63
Sullivan, Maxine, 204, 215, 220
sustainability. *See* energy technology innovations
Swan, Boots, 212
Swan, Julia, 71–72, 78, 80, 81,
Swift, Kim, 72, 73, 79, 80
Sylvania corporation, 20

T

Tatum, Art, 219
Taylor, Hosea, 213, 220
Taylor, Osie, 213
Technicolor Color Advisory Service, 176
Technicolor technology. *See also* Hollywood, California
  complications in transition to, 175
  Disney's use of, 172–180
  early laboratory of, 166
  live-action films using, 172–180
  makeup used with, 175–176
  Natalie Kalmus and, 176–177
  Process Number One, 165–166
  Process Number Two, 167–168
  Process Number Three, 168–169
  Process Number Four, 169–170
Techstars program, 200
Telluride Historical Museum, 13
Temple, Shirley, 173
Terman, Frederick, 5, 17–18, 19–20, 22, 39
Terrell, Paul, 32
Terry, Eli, 113
Thelen, Paul, 72, 74, 75, 79–80
Thomson, R. H., 225
toasting, 49
Toll of the Sea (film), 167
Tone, Tony (DJ), 48
"A Toolkit to Document Your Place of Invention," 12

transistors
  invention of, 22
  personal computers and, 35–38, 39
  used with pacemakers, 89, 101, 103
Travelers Indemnity Company, 110, 112
Tribble, Bud, 33
Troland, Leonard T., 167
True Northwest exhibition, 223
turbochargers, 194–195
turntables, 47, 52, 54, 58–60, 64
Turrentine, Stanley, 203, 208, 213, 215, 220
Turrentine, Tommy, 213, 215, 220
Twain, Mark, 110, 112, 115–116, 126, 127
Twin Cities, Minnesota, 86–87. *See also* Medical Alley, Minnesota

U

Underwood Typewriter Company, 133
United Artists, 160, 174
university-industry cooperation, 19–20
University of Alabama at Huntsville, 12
University of Colorado, 185
University of Colorado Technology Transfer Office, 200
University of Minnesota, 87, 88–98, 100, 102, 106–107, 109, 149
University of New Orleans, 13
urban planning, v–vi
U.S. Department of Agriculture's National Center for Agricultural Utilization Research (USDA-NCAUR), 12, 139, 141, 151, 153, 155
U.S. Department of Energy, 185, 197, 198
USS *Macon*, 18
U.S. Space and Rocket Center, 12

V

VanDyne, Ed, 194–195, 198
VanDyne SuperTurbo, 189–190, 194–195, 198, 199, 200
Varco, Richard, 91, 92
Varian Associates, 18, 19, 20, 23, 39
Varian, Russell, 17, 19
Varian, Sigurd, 17
Variety Club Heart Hospital, 86, 90–91, 92
Vaughan, Sarah, 206
Vehicle Technologies Office, U.S. Dept. of Energy, 198
Vesce, Sean, 70, 71, 84
video games. *See* Seattle gaming industry
Villafaña, Manuel, 106, 107–108
Volunteer Ambassador Program, 217

W

Wachtel, Jacques L., 145
Wadsworth, Jeremiah, 112
Walker, Captain Samuel, 114
Walsh, Andrew, 134
Wangensteen, Owen H., 89–90, 91–92, 95
Warner, Albert, 160
Warner, Charles Dudley, 126
Warner, Harry, 160
Warner, Jack, 171
Warner, Samuel, 160, 171
Warner, Thomas, 115
Warner Bros., 164
Washington Interactive Network (WIN), 70, 76–77
Watts, Jeff Tain, 220
Webster, Ben, 216
Weed, Theodore E., 121
Weed Sewing Machine Company, 120, 121–122, 123
welfare capitalism, 126
Wellman, William, 173
Wells, Percy, 145, 147
Wescott, W. Burton, 164, 165
Western Reserve Historical Society, 13
Westfall, Catherine, 4
Westinghouse High School, 208
Whitley, Hobart Johnstone, 156
Whitney, Amos, 119
Whitney, Eli, 113, 115
Whitney, Eli Jr., 114–115
Whitney, John Hay "Jock," 173
Wickerham, L. J., 145
Wilcox, Daeida, 156
Wilcox, Harvey, 156
Wilder, Billy, 160
Williams, David, 104
Williams, Martin, 160
Williams, Mary Lou, 203, 204, 208, 211, 214, 220
Willis, Edgar, 212
Willson, Bryan, 20, 128, 186–187, 188, 189, 190, 192, 198, 199, 201
Wilson, August, 217
Wilson, Joni, 213
winding machine, 119
Wingrove, Bob, 102
The Wizard of Oz (film), 177, 178, 179
Wolfe, Tom, 5, 25
Wolfkill, Kiki, 71
women
  in hip-hop music, 45
  in mass production of penicillin, 148–149, 151
  in Seattle gaming industry, 77
Woodward, Inc., 195

Wozniak, Steve, 9, 30–33
Wright Aeronautical Corporation, 133
WVIZ/PBS & WCPN ideastream, 13
Wyman, Jane, 162–163

X

Xerox Alto computer. *See* Alto computer

Xerox Corporation, 20, 26–29, 33
Xerox Star computer, 33

Y

Yale, Linus, 113
Young Preservationists Association of
    Pittsburgh, 13, 203–204

Z

Zanuck, Darryl, 164
zero-energy district, in Fort Collins, 190
Zuhdi, Nazih, 108
Zukor, Adolph, 160
Zulu Nation, 50–51